Antenna Towers
for Radio Amateurs

A Guide for Design, Installation
and Construction

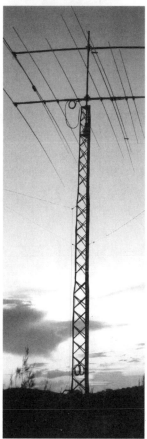

By Don Daso, K4ZA

Production Supervisor:
Shelly Bloom, WB1ENT

Layout:
Jodi Morin, KA1JPA

Editorial Assistant:
Maty Weinberg, KB1EIB

Technical Illustrations:
David Pingree, N1NAS

Cover Design:
Sue Fagan, KB1OKW

Published by
ARRL *The national association for*
AMATEUR RADIO™
225 Main Street • Newington, CT 06111-1494
www.arrl.org

Contents

Foreword

Although many amateurs rely on wire antennas hung in trees or verticals on their roofs, sometimes the urge strikes to put up something more. "Something more" might be a rotatable multiband Yagi, or a higher perch for your 2 meter vertical, or even a stacked array of monoband beams for your favorite HF or VHF bands. In any event, "something more" usually involves a tower to support your new antenna system, and that's the focus of this book.

Author Don Daso, K4ZA, presents practical information and techniques gained from decades of experience designing and building Amateur Radio towers for himself and others. A seasoned professional tower climber, Don has worked on everything from small crank-up and light duty freestanding towers to 200+ foot rotating monsters with stacks of giant Yagis.

Antenna Towers for Radio Amateurs discusses the pros and cons of various types of towers and guides the reader through the installation process from digging holes to pouring concrete to stacking sections to assembling guy wires to installing antennas and accessories. Don discusses the safety equipment, tools and skills needed for a successful project: What is a personal fall arrest system, and how do I use it? What hand tools are useful? What ropes and pulleys to buy? What specialized tools will make the job easier? Later chapters cover maintenance and insurance issues. Not everyone is willing or able to install or work on their tower, so there is a chapter describing considerations for those who decide to hire professional help.

We hope that this book will inspire you to improve your antenna system. Please work safely, follow the manufacturer's instructions, and don't hesitate to ask a knowledgeable ham or professional tower climber for help if you have any doubts or questions.

David Sumner, K1ZZ
Chief Executive Officer
Newington, Connecticut
August 2010

Preface

When the ARRL called and asked me if I would be interested in writing a book about ham radio towers, I was both flattered and worried.

Flattered to have been asked — I assumed their query was motivated by some of my earlier articles published in *QST* and my *NCJ* "Workshop Chronicles" column.

Worried that I wouldn't be able to complete the task — both because of my own expanding work schedule, and because I didn't yet know enough to create something meaningful. I find I learn something new and/or different on every tower job I do, so I wasn't sure I could cover everything in sufficient detail to meet everyone's need.

I have lots of people to thank — my parents, first of all, who let me put up my first tower on the family farm at 15. I poured the concrete one evening after my chores, then climbed up the very next day, carefully scrunched my Keds around the tower legs for security, leaned down and hauled up the next section with a short hunk of rope, lifted it overhead and into place, and then bolted it together. I did that two more times, and there it was, my first tower, upon which I later mounted a 6 meter Yagi!

I've come a ways since then. Lots of vertical miles of steel have been climbed. Lots of aluminum hung. Lots of hours spent in the air. Lots of friends, lots of clients, have all contributed to what I know today, and have tried to pass along in the pages that follow. Roger Burt, N4ZC; Lenny Chertok, W3GRF (SK); Frank Donovan, W3LPL; Hank Lonberg, KR7X; and John Crovelli, W2GD, all deserve special mention — for inspiration, perspiration, edification and motivation required to continue climbing and doing tower work! Special thanks to Kim Hinceman, K4ATX, for helping with proof-reading.

And very special thanks must go to my wife, Marti, who loves and supports me in this, the greatest of hobby adventures, and watches over my health, both mental and physical.

Don Daso, K4ZA
Charlotte, North Carolina
August 2010

Introduction

Tower... the word conjures up immediate images. Being both noun and verb, it's powerful. Thoughts of a tower can create excitement for any ham because that old urge to transmit and receive further (to work DX) is still as strong as ever and a motivating force behind many station improvements. This chapter presents an overview of Amateur Radio towers and some things to consider if you're thinking about adding one to your station.

Some History

At the close of the 19th century, America found itself replicating the European optical telegraphy system, a line-of-sight system. Poles, modest ones 20 or 30 feet in height, held flags and other signaling devices. In 1844, Samuel F. B. Morse set up a successful telegraph line of 30 foot poles alongside roads and rail lines. By the time Marconi was demonstrating his "wireless," in 1899, most major US cities were already surrounded by horizons filled with wires and poles (**Figure 1-1**).

Within the first two decades of wireless, the race to reach and receive further pushed antennas ever higher.

Amateurs, governments and seemingly every entrepreneur wanted the new wireless in some way. Along with this widespread interest came complaints that these poles and wires were bad — reducing property values, creating

Figure 1-2 — These *QST* ads from 1948 show a typical tower and rotator available to hams after WWII. Note the "Tilt-Top" mounting head to swing a Yagi boom to the vertical position for easier element adjustment.

Figure 1-1 — Even wired telegraphy required someone experienced at climbing!

interference and destroying views and peace of mind for residents. Sound familiar? Sure, that could be a lead story for your local TV news today or in your local newspaper or on the Internet.

Historically, hams have always searched for ways in which to raise their antennas ever higher. Early efforts included tall wooden poles, usually made from cedar. The metal structures we take for granted today came into more widespread popularity in the 1920s and 1930s, at first commercially. By the end of World War II, as surplus electronics became widely available, ham radio's growth followed the burgeoning technology evolution that America was undergoing. A quick glance through *QST* magazines from that period shows ads for some of the earliest commercial Yagi antennas, rotators and metal towers, along with a veritable treasure trove of war surplus items advertised for sale. **Figure 1-2** shows some examples of what was available to hams in the post-WWII era.

It's become increasingly challenging for today's ham to put up a tower, but still the need and desire for an effective antenna system persist among active amateurs. I hope that this book will give you encouragement and ideas to help overcome the challenges and make your tower a reality.

About this Book

This book will consider some of the reasons a typical ham might want or need a tower. If you take the time to answer some basic questions about what you want to do with your ham gear (the sorts of activities you enjoy and pursue), you'll be even further ahead of the game. Whether you're a DXer, or a contester, or a casual rag-chewer, a VHF operator, or even an inveterate tinkering builder, you can probably benefit in some way from having a tower to support whatever sort of antenna(s) you eventually choose to use. There are real benefits, after all, to having an antenna up high and in the clear.

Information in these chapters will help you decide whether or not it's prac-

tical to consider putting such a tower up by yourself (perhaps with the help of local amateur friends), or whether you should call upon professional help. We'll also examine some of the pros and cons to various approaches.

This volume is not intended to be a cookbook — meaning, having read it, you won't necessarily be immediately, innately qualified to buy and install your own tower. It *is* intended to guide you through some of the pitfalls and possible stumbling blocks to owning and maintaining a tower. (Maintenance remains one of the more misunderstood aspects of the typical ham tower installation.)

The following chapters will describe some professional solutions to typical ham installations, along with advantages and disadvantages to various approaches. There's a considerable amount of "hands on" information in this book, as the author has made his living working on ham radio towers and antennas for the past few years, much to his own surprise! This focus on the practical, at the expense of abstract theory, is not meant to slight or otherwise lessen its importance. Indeed, I often call on friends for help in these areas, as my degrees and experience are in the liberal arts, not engineering. The *ARRL Antenna Book* has also long been a primary guide on this incredible journey, and it's hoped that this book will serve as a useful companion piece to that volume.

Safety

Throughout this book, you'll find references and remarks pertaining to safety. In planning and performing tower work, you cannot stress safety enough! Tower work is dangerous. The simple line that appears on the little label that Rohn puts on a tower section — YOU CAN BE KILLED — cannot be taken for granted. If you're not sure of what to do, if you're frightened, if you're out of shape or otherwise physically incapable of lugging not only yourself, but also a bunch of tools and hardware up in the air, if you're not careful and concerned with safety at

every step of the process — *you can be killed.*

There's a reason tower work has been declared the most dangerous occupation in America, and it has nothing to do with your ham radio license class or experience level or attitude or physical abilities. It's that working at heights is dangerous, pure and simple. *Falling more than a few feet can kill you!*

While most everything else in ham radio can be taken for granted (only high voltages found in amplifier building or equipment repairs come close to this level of serious, potential life-threatening danger), tower work cannot be taken for granted. Everything must be evaluated, examined, and an exhaustive set of checks and balances must be met. That's necessary each and every time you consider working on a tower, regardless of whether it's 20 feet tall or 200 feet tall. And by the way, this book will assume that your ham tower will not exceed 200 feet in height. Those towers require FAA approval and must meet rather rigorous lighting and painting standards. Although there are some ham radio installations utilizing higher towers, we won't be climbing higher than 200 feet within these pages!

This level of seriousness and emphasis on safety is not meant to frighten or otherwise deter you from such work. While it's sometimes grueling labor, tower work can also be rewarding and even fun. It can provide some real highs, if you'll pardon the pun. There's a lot to be said for problem solving at heights, and there's something to be said for doing such work on our own.

Indeed, tower and antenna work represent one of the areas where the individual experimenter or builder can still shine in today's ultra-modern, high-tech, surface-mount world. And if, within these pages, you think there's a lot of emphasis placed on actually climbing — working aloft — you're right. I admit to thinking that a "real tower" is one you can climb, not merely crank up or down and tilt over, or a tubular design, and so forth. I admit to *liking* those lattice-like structures that allow you to clamber up,

Tower Height and Making Choices

Of course, one of the major factors you should consider is the height of your antenna(s). Most antennas are designed in what's known as a "free space" environment, meaning the effects of the earth are not considered during modeling. Once the design is finalized, the effects of real earth are considered. Effects of real earth vary from location to location, of course. The relatively poor conductivity of the earth absorbs RF energy, which causes loss. Antennas that are "close" to the earth (meaning less than a ¼ wavelength at their operating frequency), especially suffer from such losses.

Forgetting budgetary concerns for the moment, it's possible to have antennas that are too high, just as it is to have antennas that are too low to be effective. The old adage "if it stayed up through the winter, it's not tall enough," has been proven wrong, not only through modern modeling programs, but in practice as well. Running a program like *HFTA* (*High Frequency Terrain Analysis*), bundled with the *ARRL*

Antenna Book, is a solid recommendation for any aspiring tower builder! **Figures 1-A** and **1-B** show an *HFTA* analysis of performance of a 20 meter Yagi at 100 feet based on surrounding terrain at a specific location. Armed with results for *your* location, you can make better decisions about your tower and what antennas you might choose to put on it — decisions that will serve you well throughout your ham career.

So, how high is high enough? I get asked this question all the time, and by all types of operators. Your budget, local ordinances and homeowner's association guidelines, esthetics and what materials you have on hand are all factors that drive the answer to this question. There is no easy, simple, single answer. But here's a starting point: How seriously do you take ham radio?

I always recommend you begin simply — let experience (time) guide your decision-making once you've gotten accustomed to what you want to do with this hobby.

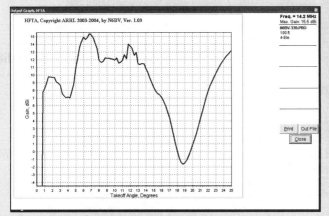

Figure 1-A — *HFTA* software, bundled with the *ARRL Antenna Book,* is a useful tool for making tower height choices based on your local terrain. This example shows antenna gain at various takeoff angles based on a 4 element 20 meter Yagi antenna 100 feet high and the terrain profile shown in Figure 1-B.

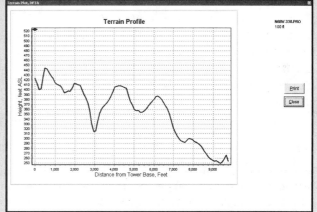

Figure 1-B — The takeoff angles shown in Figure 1-A are based on this terrain profile included on the *ARRL Antenna Book* CD. The example shows the height of terrain at a heading of 330 degrees out to a distance of 10,000 feet (nearly 2 miles) from the tower base.

wrap a lanyard around, lean back and gaze at the horizon! It never fails to please and excite me, in some small way, every time I'm "up there."

Dreaming of Towers

It's probably not too far from the truth to consider that every ham, if he or she could, would have a tower. After all, the antenna is the most important link in the communications chain, and getting it up high and in the clear is always a good thing.

In 1910, early radio pioneer Edwin

Armstrong built himself a 125-foot high antenna tower (wooden) at his family home in Yonkers, New York, while still in high school. So you can see that this urge (and need) existed from radio's very beginnings.

Such urges are always tempered by the ability to not only realize them, but to pay for them. Everything has a price. Some things that go into a tower installation that today's ham radio urge-holders should consider are listed below. If you're not familiar with some of the items, don't worry about it. We'll go into detail in the following chapters.

- the tower itself, whether new or used
- engineering drawings and/or permits
- excavation for the base and/or guy anchor holes
- rebar (reinforcing steel) for tower base and/or guy anchors
- lumber for base forms
- base section or plate
- concrete (and possible transport to get concrete to the base/anchor holes)
- guy anchors (if needed)
- guy wire (if needed)

- guy hardware (tower bracket, insulators, grips or cable clamps, shackles, thimbles, turnbuckles, equalizer plates)
- hardware required (bolts, ground rods/wire, clamps, lightning protection, etc.)
- accessories required (step bolts, rotator plate, top or bearing plate, thrust bearing etc)
- mast
- anti-climb panels
- shipping costs for material not purchased locally (this stuff is heavy!)
- crane or other heavy equipment (if needed or used)
- installation labor (one or more workers as needed, if you cannot climb yourself)
- special tools required for installation (ropes, gin pole [erection fixture], capstan winch, other climbing accessories and tools)

Note that this list is just items needed to install the tower structure itself. It does not include your antenna(s) or rotator(s), coaxial cable or control cables, or the method of securing those to the tower. And it doesn't include the method of getting cables from the radio to the tower, or providing access for them into your radio shack, or protecting them from lightning-induced dam-

Figure 1-3 — A modest guyed tower will easily hold a triband beam (20-15-10 meters) or one that includes 17 and 12 meters. This one also has a 6 meter beam at the top of the mast and supports some wire antennas for the low bands — a very effective station. The proud owner is W7VOQ.

Figure 1-5 — Is this the world's biggest ham radio tower? This massive array is part of the Radio Arcala station OH8X in Finland. It features 3 elements on 160 meters, 5 elements on 80 meters and stacked 4 elements on 40 meters. All are full size. Tower height: 100 meters. 80/160 meter antenna booms: 60 meters. (OH2BH photo)

Figure 1-4 — Taking it up several notches, K4VV's antennas are mounted on self-supporting monopole (sometimes called a "Big Bertha"). The rotator is at the bottom and the whole pole turns. This arrangement makes it easy to mount large monoband Yagis and stack them for added gain. These towers require *a lot* of concrete in the base and special equipment to erect. (K4VV photo)

age. Maintenance, which should be an annual consideration, is not mentioned.

This list is presented here, at the very beginning of this book, to demonstrate the seriousness of such an undertaking. Acquiring a suitable radio tower, and then raising that tower, is a formidable task. It's not something to be undertaken lightly, not to be considered a simple, walk-in-the-park sort of job.

This does not mean you should not consider it. This does not mean you cannot do it, working with a group of ham radio pals. It is simply intended to show the scope of the things that must be considered before you break ground, the kinds of materials needed, and the somewhat specialized system of knowledge towers require.

If you're successful, at the end of the process you might have a tower like the ones shown in **Figures 1-3** through **1-5**.

Towers, the Reality

Your radio tower represents a considerable investment, in time, money and physical labor. Done properly, your tower is a long-term investment that you can enjoy throughout your Amateur Radio career.

Towers and Local Government Regulations

Once upon a time, if you wanted a tower for your ham station, you just went ahead and put one up. That's it. End of story. But today, with the rapid proliferation of the cell phone and wireless industry across the country, things have changed dramatically. Even an installation as simple as the one shown in **Figure 1-6** requires careful planning.

Towers are sometimes, perhaps even often times, perceived — by individuals, by neighborhoods, by local governments, by companies and corporations, and even by many in the media — as negative or bad things, literal blights upon the landscape. So…what can a ham do? The answer is not as simple as we would perhaps like it to be.

You have to learn, to know, to under-stand and to appreciate (even though it may be difficult), any number of legal aspects of erecting that pride and joy in your own backyard. There are zoning ordinances and building codes to follow. You'll learn about setback requirements (distance to your property line), as well as wind and ice loading requirements. There are meetings, hearings and other hurdles you may be required to jump over. All of them are done in order to obtain a permit — which is a necessary step for every potential tower project, without exception. It's necessary so that you are protected should any "trouble" ever arise. Without a permit, you are subject not only to legal action, but also fines and the possibility of being forced to remove the tower you worked so hard to erect.

Every community (whether we're talking local, county-wide or state-wide) is different. Building permits and regulations vary widely. The best way to proceed is to actually visit the appropriate local zoning office and ask for a copy of the relevant regulations. If necessary, buy a copy — you must have them at hand for review and study. Simply asking questions, or reviewing the regulations in the office, is not enough. Not only will these regulations describe the permit application process, but the basis and process for appeals will also be spelled out (in case your application is denied initially).

Do not be surprised if you have to

Figure 1-6 — Installed in-town on a typical suburban lot, this 40 foot Rohn 25 tower holds rotary antennas for 20 through 2 meters plus wire antennas for the low bands. Although many hams would consider this a very basic installation, local building officials had not dealt with a ham radio tower before. They required a detailed design and load analysis from a registered professional engineer before issuing a building permit, a physical inspection of the base and guy anchor holes before pouring concrete, and a final inspection upon completion. (K1RO photo)

explain ham radio, ham radio towers and the like. Most zoning and regulatory offices (and officials) are used to dealing with applications for structures where folks will live or reside. A structure strictly for a hobby purpose, and not for habitation, will likely be new to them. Do not be surprised if you are required to have a registered professional engineer with knowledge of towers and antennas sign off on a detailed design and drawings for your specific tower and antenna installation (not just copies of pages from the manufacturer's catalog).

Be sure to understand and follow *all* the steps required by the building officials. If you skip some, you may find yourself presented with a "stop work" order while you resolve outstanding issues. Worst case, you may be forced to re-do work with the paperwork and inspections.

PRB — How Does It Apply?

The FCC has an interest in ensuring that the stations it licenses can have antennas that are appropriate for the services they are expected to render. In the case of the Amateur Radio service, the FCC formally expressed that interest in 1985 in an FCC declaration of limited preemption of state and local regulations. Commonly known as "PRB-1," the FCC declaration says that local governments must reasonably accommodate Amateur Radio operations, but they may still zone for height, safety and aesthetic concerns. Full details may be found at on the ARRL Web site.

PRB-1 does not specify a minimum height below which local governments cannot regulate, but they must be "reasonable" in their accommodations.

Some states have adopted state statutes, which codify PRB-1. Five of those states — Alaska, Wyoming, Virginia, North Carolina and Oregon — specify heights below which local governments in those states may not regulate. This actually goes further than PRB-1. Amateurs in other states are trying to convince their legislators to sponsor similar legislation.

The following states have adopted state statutes: Alaska, California, Florida, Idaho, Indiana, Kansas, Louisiana, Maine, Massachusetts, Mississippi, Missouri, Nevada, New Hampshire, New Mexico, North Carolina, Oklahoma, Oregon, Pennsylvania, Tennessee, Texas, Utah, Vermont, Virginia, Washington, West Virginia, Wisconsin and Wyoming. This list may have changed since I prepared this book, so check the ARRL's PRB-1 Web page for the latest details.

Organizations Against Towers

I was surprised, when researching various topics for this book, to encounter so many *organized*, apparently well-funded (lots of lawyers represented), and well-written treatises *against* towers. Granted, much of the focus is on the burgeoning cell tower growth across the US, but the point is, there's a *lot* of resistance out there and it's growing! This is something you need to be prepared for as you lay the groundwork for your tower project. For example, Scenic America's booklet, *Taming Wireless Telecommunications Towers*, remains one of the best overall arguments against radio towers. Learning about the common objections is useful because you need to be prepared to address them when making your case.

Resources

Tower manufacturers publish a lot of technical information about installing and using their products. Your vendor may be able to supply enough design information and drawings to help you satisfy your local building department. As mentioned previously, it's common for building officials to require that a Professional Engineer licensed in your state sign off on your design. If that's the case, other hams in your area may have a recommendation. ARRL sponsors the Volunteer Consulting Engineer (VCE) program and may list someone in your area. VCEs are familiar with ham radio and with typical tower installations, so they'll know about and have a good idea of what you are hoping to accomplish. See the ARRL Web site for details.

ARRL publishes quite a bit of information that may be useful in developing your tower permit application on the ARRL Web site. In addition, the ARRL publishes *Antenna Zoning for the Radio Amateur* by Fred Hopengarten, K1VR. The second edition, revised and expanded, had just been released when I wrote this chapter. Fred's book is simply invaluable in helping you navigate through the red tape! Another valuable resource is K1VR's Web site, **www.antenna zoning.com.**

If you encounter antenna restrictions or zoning issues that require professional legal assistance, an ARRL Volunteer Counsel may be able to help. Volunteer Counsels (VCs) are attorneys who are familiar with Amateur Radio and antenna restrictions. VCs will provide an initial consultation free of charge, but expect to pay a fee if you hire one to represent you. As with Volunteer Consulting Engineers, the advantage is that VCs are familiar with ham radio and tower regulations, so you won't have to spend valuable time bringing them up to speed. For more information and a list of VCs, see the ARRL Web site.

TIA/EIA Standard

There is an industry standard that applies to commercial tower construction and manufacture. Introduced in 1949, it's currently published by the Telecommunications Industry Association, although it was originally published by the Electronic Industries Association — which is how it came to be referred to as the TIA/EIA standard. Hams and their attendant tower manufacturers do not have to follow TIA/EIA guidelines, but there's a wealth of information, including design data, that any enterprising tower builder (or potential builder) would be wise to consult and consider.

The current version (or revision, if you will) of this standard is Revision G, which became effective in January 2006. There were significant changes in this latest release, one example being a more international focus. And another change was the shift away from Allowable Stress Design (ASD), the prevailing methodology used in the US for many years, to Load Resistance Factor Design (LRFD), which is widely accepted as more accurate.

Revision G is twice the size of Revision F, comprising 15 chapters, supported by 14 annexes, providing users with the minimum criteria for specifying and choosing steel antenna towers. Wind loads, as well as ice loads, are dramatically different from those used in previous publications. Wind used to be considered in terms of "fastest mile per hour," and now it's averaged as a three-second gust. Ice used to be considered to be the same everywhere, and now it's known (and shown) to be different for different locations. Detailed maps illustrate ice and wind conditions throughout the USA and its territories. Another change is the acknowledgment that location determines how towers are designed, built and erected. Previously, towers were simply built to the same standards, regardless of location. Revision G introduced a classification system for towers that depends upon their proximity to people and the services that tower supports.

The three classes of towers found in this Revision include:

Class I: Structures that due to height, use, and/or location present a low risk to human life and property damage in the event of failure, or that are used for services that can be considered optional, or where a delay in returning such services to operation would be acceptable. Such services might include residential wireless, TV, radio or scanner reception, wireless cable service, and Amateur Radio.

Class II: Structures that due to height, use, and/or location present a substantial risk to human life and property damage in the event of failure, and are used for services that may be provided by other means. Such services might include commercial wireless communications, TV and radio broadcasting, cellular and personal communication services, CATV and microwave communications.

Class III: Structures that due to height, use, and/or location present a high risk to human life and/or damage to property in the event of failure and that are used primarily for essential communications. Such services include communications supporting civil or national defense, emergency, rescue or disaster communications, and military and navigational services

Revision G is the first time Amateur Radio services have been considered within the TIA/EIA standard. Despite being ranked as the lowest class of service, there are a variety of issues any ham even considering having a tower should find of value.

Such topics include:

- tower loading and stresses
- foundations
- tower anchors
- guying
- grounding
- maintenance
- inspections
- ice loading (updated in Revision G to include every US county)
- minimum wind speed ratings (for every US county)

Obviously, the first thing any ham should consider for his/her installation is the last listed item — the wind speed rating within his or her locale. You then read the manufacturer's guidelines published for the tower you're considering or intending to use, along with the appropriate load you intend to put on that tower.

More information on the TIA/EIA standard may be found in the **Appendix** to this book.

Some Things I've Learned Along the Way

Having a fair amount of experience in the field of ham radio tower and antenna work, I'd be remiss if I did not share some of the more common tower building mistakes encountered throughout the years of doing such work. No names or call signs will be presented here (to protect the guilty parties), but all these generalizations were derived from actual case histories! These are general ideas that you should keep in mind throughout the process of planning your tower, assembling the pieces, and putting everything together.

The Two Biggest Mistakes

The single biggest error encountered is *not following directions*. Not doing what the manufacturer wants you to do ranks as the number one mistake on the part of builders, both newly licensed and experienced hams, probably now and forever. It's simple: *read and follow the manufacturer's directions*!

The second biggest mistake is *not inspecting the installation* — either often enough, or sometimes *never*! I've seen Rohn bases (even ones built above grade) where two of the three tower legs had rusted totally through. While that's a critical (and dangerous) situation, other examples abound: coax connectors unprotected or about to fail, rotation loops that have come undone, coax runs that have come apart in crank up systems ... the list goes on. You simply cannot have a mechanical/electrical "system" such as a tower and antenna installation, which you do not inspect with some regularity. Otherwise, you're simply living on borrowed time until the first, or the next, failure. And we all know that Murphy's Law postulates such failures will occur at the most inopportune time. We'll deal with inspections in some detail in the chapter on maintenance.

Other Factors

Other mistakes are difficult to rank, in terms of something like priority or order. But here they are:

Improper guy tension. In over 20 years of tower jobs, I've encountered exactly one tower installation where the guys were "too tight." In all other instances, they were simply way too loose. You could practically see them oscillate! This allows the tower, the guys themselves, and especially the load at the top, to move excessively.

It's an easy fix, covered in detail in a later chapter.

Overloading. Hams do this all the time. And they often get lucky. Manufacturers build in some security, for obvious, potential litigation issues. Sometimes, too, the actual tower is stronger than specified. Sometimes, too, the location shields the tower from damaging winds, and so forth. But just don't do it — keep in mind the single biggest mistake, previously mentioned. And keep in mind what the intended load should be.

Choosing an improper mast. If I had a dollar for every piece of water pipe I've replaced ... well, you know the rest — I'd be living in the Caribbean, hamming up a storm. The yield moment of a somewhat typical installation, say a two element shortened 40 meter beam mounted 10 to 12 feet above a tribander, can put tremendous stresses on your mast. So, it stands to reason that water pipe made from unknown materials, intended only to transport liquids, is *not* what we want up there on the tower. Do the calculations, talk to your local steel supplier or tower hardware supplier — whatever it takes — and install a proper, suitable strength mast.

Failure to install a proper ground system. This is a necessary item in order to protect your home from lightning strikes, of course. But it can also prevent RF interference. Entire books have been written and devoted to this topic. We'll cover it in detail in the chapter on bases.

Hardware issues. This final category refers to not using "the right stuff" in your tower and antenna work. Stainless steel and/or galvanized fasteners are really the only choice when it comes to longevity when confronting Mother Nature. Knowing what hardware to use, when, and how to install it all correctly ... all covered in detail in later chapters.

Basic Tower Types

There are two broad types of tower from which you can choose: *guyed towers* and *self-supporting towers*. The category of self-supporting towers, also referred to as *freestanding*, is also broken down into two classes: crank-up tower models that can be raised and lowered, and the more classic, tapered self-supporting models at fixed heights. Each of these versions can be constructed from aluminum or steel.

There are only three major considerations in choosing the correct tower style for your station: load requirement, tower footprint and cost.

First, you need to determine the antenna load requirements of the installation. This is a function of the antenna's projected wind area, along with all the coaxial cables and control lines, brackets, rotators and the like that are mounted on your tower. All towers have load ratings (in square feet) under various conditions of wind speed and ice load. You must know how much load your planned antennas present to select an appropriate tower and install it properly. It's a good idea to account for possible changes in your antenna system later on so you have the flexibility to add antennas or put up larger ones as your interests evolve.

For example, if you're newly licensed, you probably do not yet know exactly all of the areas that interest you in the hobby. A tower that's tall enough

and strong enough to support HF antennas is considerably more effort and expense than one to support a few VHF/UHF antennas. If you're just starting out on VHF, but plan to add HF capability at some point, it's easier to plan for that now — rather than putting up a light duty tower and then dismantling and replacing it with a bigger one in a year or two.

Second, you need to determine the tower footprint; that's the amount of land required for your installation. Considerations include the location of trees and buildings that could interfere with the construction of the tower and installation of antennas; property line or building setback requirements imposed by local regulation; placement of guy wires (if needed); and aesthetic concerns expressed by family members who may not see the same beauty in the tower as you do. For reasons discussed in the following sections, the smaller the footprint, the more costly the tower itself and installation will usually be. Freestanding towers have the smallest footprint, but they use more steel in their construction and require a more substantial concrete base.

Speaking of costs, that's the third and final consideration. What will your budget allow?

Yes, height is, of course, also a factor to consider. As explained in the sidebar in Chapter 1, I suggest running an *HFTA*

(*High Frequency Terrain Analysis*) prediction for your location before starting construction. The analysis includes those bands that hold your interest and the antenna(s) you're considering, and in a perfect world will help you determine optimum antenna heights.

An *HFTA* analysis works very well, but sometimes that's not the yardstick the station owner is using. For example, let's say you have a 70 foot tower available, so that's what you're going to put up regardless of what the software predicts. Or perhaps you have limited space or other requirements that keep your maximum to 40 feet. That's fine — go ahead and plan to put up what tower you can. Get your antenna up in the air, and enjoy operating.

Freestanding Towers

The amount of concrete used in the tower's base is related to how far apart the legs are spaced, along with the overall height of the tower. Freestanding towers with their legs set relatively close together require larger concrete bases, along with larger steel. Guyed towers, in contrast, have geometry working in their favor, with the wide spacing of the guy wires.

The end result of all this (simplistic) physics is that guyed towers require less steel in their structure and less concrete in the ground. Freestanding

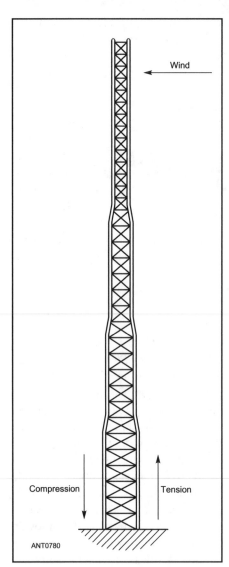

Figure 2-1 — A typical freestanding tower tapers as it rises. The wide stance at the bottom ensures stability.

towers require significantly more steel and concrete. This can mean a more expensive tower, as well as higher installation costs.

There are two types of freestanding tower — the monopole and the lattice tower. Monopoles are structural tubes, tapering from large at the base to smaller at the top. (Rohn's monopole takes up only three square feet at the base, while still rising to 100 feet.) Monopoles can be aesthetically pleasing, but they use more steel and concrete compared to any other tower. As such, they are relatively rare among amateur installations, although they are very common in cell phone installations.

Freestanding lattice towers (either three or four-legged varieties) are more common. A typical example is shown in **Figure 2-1**. They take up relatively little ground space, which is their primary attraction to most users. Some people are simply turned off by guy wires or would have to install the tower in a location where guy wires would interfere with household activities.

The crank-up variety of freestanding tower will be discussed separately.

BX Series Towers

For many hams, the BX series of self-supporting tower represents a rather low cost introduction to ham towers. Indeed, many people start out with this class of tower, and the one shown in **Figure 2-2** will provide many years of enjoyment for its owner. The BX tower (referred to as the Standard model) is the simplest version of this series. The HBX (referred to as Heavy Duty) is a heavier-duty version, and the HDBX (referred to as Extra Heavy Duty) is even heavier-duty. Manufactured for many years by Rohn, these towers are now made by Thomas Shelby & Company and are still widely available to hams.

All versions of BX series tower use sections of various sizes and configurations, all eight feet in length, to achieve the desired height. All versions use X-bracing. The legs are neither round, tubular members nor angle stock, but rather use a stamped, channel steel style construction. The X-braces are riveted together, and then riveted to the channel legs. This construction, combined with the relatively mild steel, makes them rather lightweight towers. Again, consider them as introductory, "first tower" models, and you'll appreciate not only their economy, but also their functionality. When disassembled, each tower section telescopes inside the next larger section, making them easy to transport.

The X-braced legs can be difficult to climb. They are certainly difficult to stand on for extended periods of time, especially at the top, on the smallest section, which is exactly where you'll be working for long periods! Since the legs are stamped steel, they will not withstand the torsional forces of a more

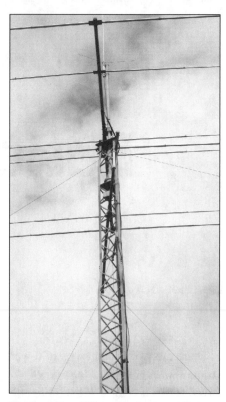

Figure 2-2 — BX series towers offer an economical choice for small HF and VHF/UHF antennas. Typical installations tend to be 30 to 60 feet high. Here, N4PQX's HBX-56 tower supports a Force-12 C3 Yagi for 20 through 10 meters and his 30 meter loop.

typical round leg tower. And, since the bracing is riveted, and not welded, the sections are not as strong.

The small size of the topmost section can make inserting a rotator somewhat difficult. For example, the popular Hy-Gain T2X Tailtwister model won't fit inside the top section at all, although I have had good luck in using a long mast, with the rotator mounted well down inside the tower in models from the light-duty BX-40 (40 feet) to the heavier HBX 56-footer.

None of these points should discourage you from using or putting up one of these towers. They are inexpensive, unobtrusive (especially if painted flat black!), and more importantly, they can be found seemingly everywhere.

Aluminum Towers

Self-supporting towers constructed of aluminum are also available. These towers are made by the Heights and Universal companies, and they are available in various configurations and heights. Popular because of their light weight, aluminum towers are simply not as stiff as steel towers, and will not carry similar loads. Consider them for VHF antennas or smaller HF Yagis.

Special mention should also be made of some products offered by the Glen Martin Engineering Company — specifically, their all aluminum-angle, bolt-together design tower. They also offer a "Hazer" system, which is a tracking device designed to allow you to take an antenna up and down 50 feet of Rohn 25G without climbing.

Heavier Duty Freestanding Towers

Say you want a tower that can handle larger antennas or is taller than the 40 to 60 feet typical of the light duty BX or aluminum freestanding towers. If a guyed tower is not an option, a heavy duty freestanding tower is another consideration. Heavy duty towers from AN Wireless, Trylon and Rohn are also readily available to hams. These are tapered designs (going from larger base sections to smaller top sections) available in a wide range of heights and load ratings. They offer some distinct advantages.

AN WIRELESS TOWERS

Let's first consider the AN Wireless (ANW) towers. All ANW towers use 50 ksi steel throughout the entire structure (ksi is a rating for kilo-pounds force per square inch; 1 ksi = 1000 psi). ANW uses this steel for added strength and weight saving features instead of the typical 36 ksi material that's found in many tower designs. The towers receive full hot-dip galvanizing treatment after fabrication to prevent rust on the edges or at joints. ANW towers provide significant wind loading specifications, including a tower and foundation with a sustained 120 MPH wind zone rating.

With 24 bolts at each section-to-section intersection, along with heavy steel plates on each side of the joint for support, this is a rugged tower. Bolt diameters are the same at the top of the tower as they are at the bottom, too. The tower is strong enough to be fully assembled, then lifted into place by crane without worrying about stressing the joints. ANW towers don't lose height during construction, since there are no overlapping joints — a 100-foot tower is 100 feet tall. ANW offers optional ⅝-inch step bolts, spaced every 18 inches, so you're not climbing braces installed at awkward angles. **Figure 2-3** shows a 60 foot ANW tower under construction.

The standard foundation includes a half-section of the next largest tower section, complete with horizontal bracing that's embedded in concrete for added strength. Custom hardware such as mounting plates, beacon plates, side arms, work platforms, rooftop mounting kits, ice shields and waveguide bridges can be built to order. Spare parts are readily available, and are field replaceable using just a wrench. Fall arrest options are available for one face, or all three.

Figure 2-4 shows a 90 foot AN Wireless freestanding tower with two large and heavy multiband Yagis. Clearly, a heavy duty freestanding tower can offer height and antenna load capabilities that rival guyed towers.

TRYLON TOWERS

Another supplier of heavy duty freestanding towers is Trylon TSF, a Canadian-based company. They began in 1932, manufacturing wind turbine generators and the towers required to support them. World War II saw a huge

Figure 2-3 — Here's a 60 foot AN Wireless tower under construction.

Figure 2-4 — This heavy duty 90 foot tower from AN Wireless is capable of supporting some big hardware without the need for guy wires. (N3KS photo)

growth for the company, with the redesign of their towers for communications use by the US Signal Corps.

Today, Trylon makes and markets a wide range of towers, but is perhaps best known among radio amateurs for their freestanding Titan self-supporting towers. The Titan towers occupy a minimal footprint area and come in heights ranging from 16 to 96 feet. Their Super Titan models continue upward in height and heavier load-carrying capabilities.

The Titan towers offer excellent flexibility for a wide range of applications. Models are available supporting up to 300 pounds, and a maximum antenna area rating of 99 square feet. The towers are rated to 100 MPH. They're usually in-stock at authorized distributors, providing fast lead times and quality product in one tidy package.

ROHN SSV

Rohn's SSV towers are designed for broadcast, commercial and industrial clients needing towers up to 500 feet tall. Obviously, an SSV tower is probably going to be capable of handling any amateur's need! Even the most basic model is designed to handle 30 square feet of antenna area, so higher loads are no problem. The knock-down construction allows on-site assembly, which can reduce transportation costs.

SSV towers are intended for commercial use, so there are tradeoffs for ham use. Namely, the "typical" ham hardware we associate with our towers and antennas (a rotator mounting shelf, for example) will not be available. Some custom fabrication is usually required. Their massive size and continual taper can make side-mounting

antennas difficult, although straight, non-tapered sections are also available. That massive size means a sizable investment in concrete in the ground!

Because SSV towers have been around for a while and used extensively in the broadcast and commercial arenas, it is sometimes possible to find them for sale or auction at what appear to be astonishingly low prices. Of course, you do have the problem of transporting them or even sometimes actually taking such a behemoth down! Once you get your treasure home, there's the considerable angst and anguish required to re-erect it, again not a trivial task. Yet, if you have a unique and demanding location (say, on a mountaintop or ridgeline), these babies can handle the wind and ice loads all day long without worry or question.

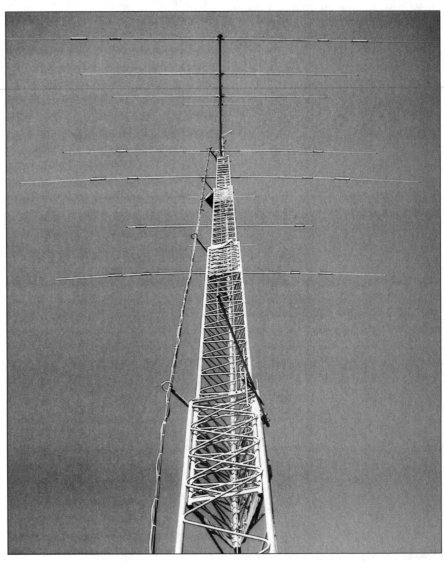

Figure 2-5 — Two examples of crank-up towers. The tower at W8DX (above) is a monopole made from telescoping tubes, while the tower at KX8D (right) uses lattice-style sections that nest inside each other when retracted.

Crank-Up Towers

Crank-up towers are the double-edged sword of the tower world — their installation and use represent real consequences for the owner. Here's why.

Not everyone resides in the country or on a suburban lot large enough to allow a guyed tower or a large, self-supporting tower. Most people do have the room to allow installation of a crank-up tower. Crank-ups are typically lattice construction, although there are tubular designs as well. The principle and operation are the same for each.

Both kinds use sections that telescope (**Figure 2-5**). Multiple sections of tower are nested, one inside another, and are pulled up and held in place by the crank-up cable. Some crank-up models use multiple cables (often called *positive pull down* design) that literally haul a section down as well as up. They don't simply rely on gravity for downward movement. See **Figure 2-6**.

It should be apparent (having read this far) that such mechanized designs are limited in their load-carrying ability. The entire tower (along with all the rotator, mast, cables and antennas) will be carried, moved and held by that single cable alone. By design, the telescoping crank-up tower sections use ever-decreasing sizes of lattice tower or tubular steel. The smaller sections near the top of most crank-up towers mean lower load-carrying capacities than fixed towers. Still, their limited footprint, lack of guy wires and ability to be lowered when not in use (presumably less noticeable to neighbors) make crank-up towers a popular choice for many hams.

Special Precautions

It's easy to simply assume that the crank up will solve many problems, but you must take some special precautions into consideration. For example:

■ Delivery and off-loading of the tower at your location.
■ Base and concrete requirements.
■ Antenna and tower maintenance.

Let's take a closer look.

It's all too easy to forget that the crank up tower will arrive at your house fully assembled — all the sections will be telescoped together, with appropriate sheaves, bearings and cables already assembled. The tower will weigh from several hundred to perhaps a couple of thousand pounds! Unless other arrangements have been made, you will be responsible for unloading that heavy, awkward tower from the shipper's truck. When the driver calls or knocks on your door is *not* the time to be thinking about how you'll unload your new tower. You must plan ahead and be prepared.

I've used everything from a crew of friends to all-terrain forklifts to automotive wreckers to A-frames and chain hoists to deal with taking a crank-up tower off a truck. Planning and preparation are critical factors to avoid damage not only to the tower, but also to you, your helpers or your property.

It's wise to remember that using the tower vendor's recommended commercial motor freight company for shipping provides you with some insurance, should the unthinkable happen and your tower suffer damage in transit. If you've arranged pickup or provided for your own shipping, you're usually on your own if something goes wrong.

When you're dealing with such large, heavy objects and the equipment required to lift and move them, it's also wise to keep in mind where that equipment is going to have to travel. Are there sprinkler heads embedded in the lawn, for example? They won't survive being run over. Is the lawn itself capable of supporting vehicular traffic? Again, pre-planning is crucial in this regard.

A crank-up tower derives its strength from a large chunk of concrete (the tower base) in the ground. Of course, crank up towers require *large* concrete foundation bases. No compromises are possible. The size of the base and the amount of reinforcing steel required, along with the concrete (and the finishing it requires) must all follow the manufacturer's guidelines to the letter. If you have any questions or doubts about your ability to do that, hire it done professionally. We'll discuss building a base and working with concrete in detail in a later chapter.

While it may seem unreasonable to talk about maintenance — even *before* the tower is purchased, let alone erected — it's more important and necessary with a crank-up than with other designs. The number of moving parts alone dictate this degree of attention. Many hams choose a crank-up, thinking that the nested position will provide easy access to their antenna system, or allow an added degree of safety during approaching storms, but crank-ups have special issues that must be taken into account.

Figure 2-6 — This crank-up tower is fitted with a "positive pull-down" to lower sections, rather than relying on gravity.

Crank-Up Towers: Things to Consider

Crank-ups offer the advantage of being able to lower the antenna(s) when the weather is threatening. Some suggest that the ability to work on the antenna once it's lowered is also an advantage, but most crank-ups leave the antenna resting at 20 feet or more above the ground — not truly convenient.

Physically cranking the tower up and down can sometimes be problematic. It takes a *lot* of energy to crank by hand, so motorized towers have come to be more and more popular. But with this simple elegance (push a button, watch your pride and joy soar upward or return to earth), comes a hefty price tag. And despite the urge, I suggest you only do the raising and/or lowering when you can actually watch the tower moving!

One solution to the requirement that you climb up to work on the antenna is allowing the crank-up tower to tilt over. Again, this clever solution comes with a significant price tag. The forces encountered while raising and lowering the tower (even cranked down) with antennas mounted on it can be tremendous.

As discussed in Chapter 1, the most troublesome aspect of installing a new tower these days can be dealing with bureaucratic issues and securing the required permit. This process can be particularly difficult with crank-up towers, since most of them are rated at 50 to 60 MPH wind loads. Building codes usually require higher ratings, so permits may be difficult to obtain without careful planning. How much load will the tower carry at the wind speed rating for your area? Even if you do get a drawing from the manufacturer, getting a needed analysis and approval from a Professional Engineer (PE) licensed in your state can sometimes be difficult. Some manufacturers are able to provide a design that will meet your needs and can help with the PE stamp of approval.

Keep in mind that load specifications are normally given for a fully extended (cranked up) tower. Variables include the wind speed, the height, the load and the bending moment. Leaving the tower cranked down (sections nested inside each other), even a small amount, will increase the wind load capabilities considerably. Again, check with the manufacturer regarding your needs.

Working on Crank-Up Towers

Working on crank-up towers can be dangerous. *Never* climb a crank-up tower that's extended. **Figure 2-7**

shows a crank-up tower that collapsed after a cable failure. Nobody was injured, but imagine if someone had been on the tower at the time!

If telescoping sections bind or hardware gets caught, a crank-up can appear to be fully nested when in fact it's not. There are horror stories of fingers and toes being jammed between sections as they slid downward because not all sections of the tower had nested.

The solution to this problem is to insert a section of pipe or sturdy material between each of the nested sections, near the bottom of the tower, before you even *think* about climbing the nested tower to do anything. Use as large a piece of pipe as can be maneuvered between the sections. If the tower is motorized, remove all power from the motor, either by flipping the breaker off or by removing the power leads from the motor itself. Then, and only then, should you consider climbing the tower. **Figure 2-8** shows a climber doing it the right way.

Climbing those nested sections will be difficult, since there's now precious little space for your feet and hands! All these cautionary tales and troublesome

Figure 2-8 — John Crovelli, W2GD, working atop a crank-up tower. The tower is fully nested and unseen, at the base, are two pieces of two-inch steel to prevent any section from sliding or moving. Always securely block the sections if you must climb a nested crank-up tower.

Figure 2-7 — What can happen if/when your crank-up tower's cable breaks? In this case, the mounted antennas sustained significant damage as everything came to a sudden stop. (K9ZW photo)

aspects are another argument for using a simple bucket truck or a manlift to get you the 20 or so feet in the air necessary to work on the antennas or rotator. Manlifts make good sense, and they're not only cheap to rent, they're easy to operate. I consider it easy insurance for this sort of tower work. Manlifts are discussed in more detail in a later chapter.

Maintenance

Maintenance on crank-ups (since they are moving, mechanical systems) consists of checking oil levels in the motor's transmission or gearbox, greasing any bearings or fittings you see, and examining the cable for nicks, cuts, loose strands and the like. Most crank-ups use either stainless or galvanized aircraft cable (7×19 strand, most commonly), and you'll easily see if your cable is damaged. You should mark the pulleys on your tower. A simple painted-on stripe will allow you to see if they're turning, even from a distance. Watch for smooth, free rotation as you raise and lower the tower.

That cable will probably need replacing at some point, and while you *can* do it vertically, it's more easily done with the tower resting on the ground. That means everything has to be dismantled, and this is not a job to be taken lightly. Some manufacturers are reluctant to provide the required cable sizes and specifications or directions on how to re-string them — apparently because of litigation concerns. Seek professional help if you must perform this maintenance.

Greasing the cables is always a topic that generates considerable conversation. Some suggest the grease attracts dirt and dust, which can then cause abnormal wear. I usually leave stainless cables alone, choosing to lubricate only the galvanized versions. US Tower recommends replacement after three years. At this time, they are the only manufacturer I'm aware of that publicly recommends cable replacement.

Some crank-up tower vendors offer a raising fixture, which will allow the tower to tilt over to provide easier access (**Figure 2-9**). Still, you're likely to be working from a ladder at some 30-degrees or so of angle. Even though

Figure 2-9 — Some crank-up towers have a hinged base and erection fixture to bring the heavy tower from horizontal to vertical. (KX8D photo)

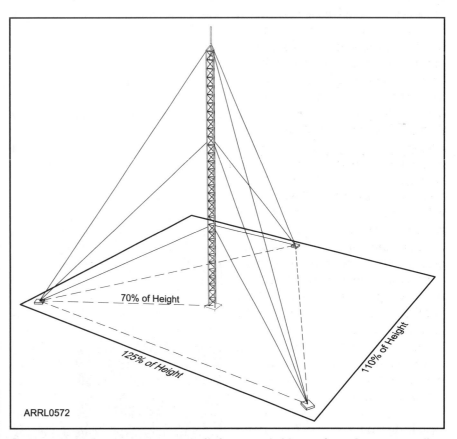

ARRL0572

70% of Height

125% of Height

110% of Height

Figure 2-10 — Guyed towers are built from stackable sections that are usually identical except for the base that attaches to the concrete and the top section that supports the mast. Guy requirements are specified by the manufacturer, but spacing the anchors 70 to 80% of the tower height is typical. As seen in the drawing, the tower and guy wires do take up quite a bit of room.

the raising fixture is rated to hold the tower, it's always a good idea to put some sort of stand or jack under the tower to take the weight while you're working on things.

Guyed Towers

Guyed towers offer the widest variety of configurations. They are built from sections (typically 10 feet long) and can be anywhere from a modest 30 or 40 feet to several hundred feet in height. They require less concrete in the base than freestanding towers, but they do require some room for guy wires as shown in **Figure 2-10**.

Rohn's G-model series of towers — Rohn 20G, 25G, 45G, 55G and 65G — represent probably the most widely used tower in ham radio circles, with 25G and 45G the most popular. (And yes, there is a 35G, sometimes referred to as "Motorola" tower, since it was manufactured under contract to Motorola; 35G is halfway between 25 and 45, but is rarely seen or used.) Each triangular section is 10 feet in length. Sections bolt together with two bolts per leg, simply stacking one above the other. A wide variety of accessories are available. Various style top sections, plates, accessory shelves, guy brackets, base options, and so forth provide extreme versatility — indeed, it's perhaps the most versatile tower system ever engineered.

G series towers are designed for efficiency, strength and versatility. They are entirely welded and fabricated with precision equipment and are suited to meet a wide variety of loads and needs. All the towers in the G series are constructed with high strength steel tubing for the legs and feature Rohn's exclusive Zigzag solid rod bracing to provide exceptional strength. They are hot dip galvanized *after fabrication*. In this process, each section of the tower is totally immersed in molten zinc, allowing every square inch of the tower, inside and out, to be completely covered. Hot dip galvanizing protects all points of welding and construction against rust and corrosion, while providing an attractive finish.

The Rohn catalog, available on Rohn's Web site, represents a virtual "Bible" of tower design data and layout and load considerations. For that reason alone, it's worth having. If you intend to put up any of the G series towers, it's necessary to at least consult the factory specifications to ensure you are doing what the manufacturer intended. That should always be your mantra in any tower construction project. Quite simply, I consider this manual invaluable. As seen in **Figure 2-11**, guyed towers can be designed to carry incredible loads.

Here's a breakdown of some popular choices from Rohn.

Rohn 25G Tower. Rohn 25G is built on a 12 inch equilateral triangle design with 1¼ inch OD legs. Sections are available in the standard 10 foot length and a 7 foot length, which is UPS shippable. The 25G tower can be used in guyed, self-supporting or bracketed configurations according to specifications in the Rohn catalog. As a guyed structure, the 25G can rise to a maximum of 190 feet. Self-supporting and bracketed heights depend on loading.

Rohn 45G Tower. 45G provides excellent strength for applications up to 300 feet. It's offered with either heavy steel tube or solid steel rod legs to satisfy a wide variety of needs under varied conditions. When properly installed, the standard tower will support loads as shown on various guyed and self-supporting information sheets. Built on an 18 inch equilateral triangle design with 1¼ inch OD legs, Rohn 45G is available in 10 foot sections.

Rohn 55G Tower. Because of its rugged design, 55G lends itself to a wide variety of uses within the communications field, particularly where unusual wind loading and height requirements exist. The 55G was designed to provide excellent strength in heights up to 400 feet. When properly installed, the standard tower will support loads as shown in the catalog on various guyed and self-supporting information sheets. Like 45G, 55G is built on an 18 inch equilateral triangle design and comes in 10 foot sections,

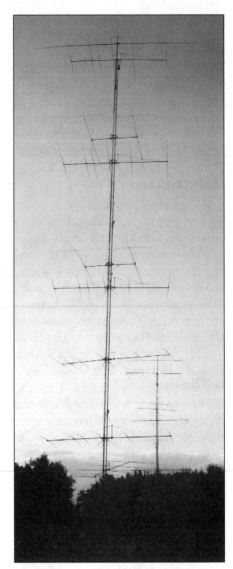

Figure 2-11 — Guyed towers can be designed to carry massive antenna loads. This one has a stack of four 6 element 20 meter Yagis, a pair of small 40 meter beams, a pair of 7 element 15 meter beams and a rotatable 75/80 meter dipole at the top. (K3LR photo)

but the legs are larger — 1½ inches OD.

Some Simple Tower Physics

The physics of radio towers is not that complex, but a simple, yet solid understanding of what's "going on" is necessary if your installation is to be successful. That old joke concerning towers and antennas that goes "If it didn't fall down, then it wasn't big enough," while humorous, truly doesn't apply in this day and age of using a solid, systems engineering approach to the design, construction and installation of a traditional ham radio tower. We want to put things up so that they survive. We put them up to stay up.

First, let's consider the tower itself — the actual physical structure. It must be strong enough, in and of itself, to withstand the wind forces that will want to cause it to bend (what we call *bending moment*). The tower base (where and how it is attached to the ground) will have to be large enough so that the tower will not tilt or fall over as a result of those bending moment forces.

The stability of any tower is a function of the spread of its legs. Think of something as simple as a camera tripod. Bringing the legs closer together causes the tripod to become less stable. Spread the legs out and the tripod becomes more stable. With freestanding towers, the three legs (wider at the base)

are analogous to the tripod legs. With guyed towers, the guy wires are analogous to the tripod legs.

Guyed Towers vs Freestanding Towers

Clients, or potential clients, or folks who are simply considering towers, often ask me point blank: "What's best, and why?" or, "What's the strongest tower I can put up?" The bottom line is that guyed towers have larger antenna area support capacity than do freestanding (cantilevered, or self-supporting) towers having similar sectional properties. Now…here's why — from a professional engineering perspective, courtesy of Hank Lonberg, KR7X:

This increased capacity comes from the way in which lateral loads from the wind are conveyed to the ground. Structural engineers call this *load path*. This mechanism of load transfer should be continuous, and the more direct it

is, the better.

Looking at **Figure 2-12**, we see that on the freestanding tower, the tower section itself is the principal load path to the ground. In the guyed tower example, the principal load path for the lateral load will be the individual guy wires. The tower section provides a load path for the lateral load it develops from the wind on its surface. In addition, the vertical component of the guy force, developed from the resistance to the lateral loading, will have its load path to the ground through the tower in axial force.

In the case of a freestanding tower, the forces developed at the base are axial force (P), lateral shear force (V) and bending moment (M). These are resisted by the tower section, and in turn resisted by the foundation or base structure. If they were not, then the freestanding system would not be in equilibrium. It would go sailing off into space. Sir Isaac Newton developed the

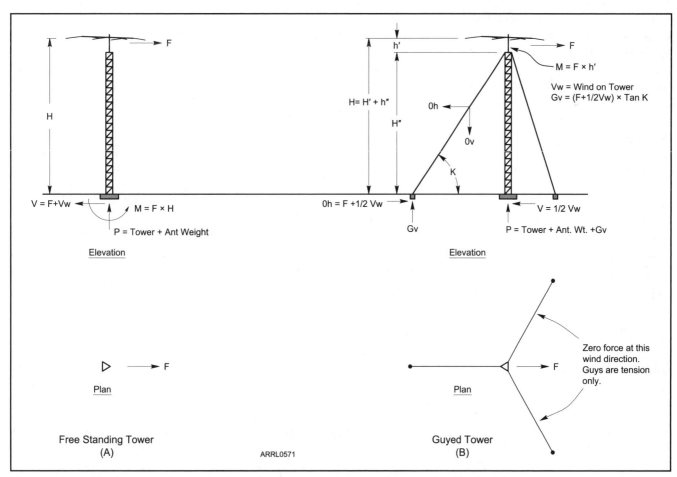

Figure 2-12 — The forces acting on freestanding and guyed towers. See text for discussion.

mathematics and theorems for determining these forces.

Looking at the guyed tower sketch, you will see that the lateral load from the antenna (above the guy point), is transferred to the ground by the guy wires. This load is transferred down to the guy point by the tower and the mast, which is attached to the tower. Below the guy point, the tower only sees the lateral load developed by the wind — not the lateral force developed by the antenna (which is usually much larger than the wind load on the tower). Each guy, if activated by the force direction, develops a horizontal resistance to the lateral load and also develops a vertical component based on the geometric arrangement of the guy in space. The axial load on the guy in tension is the sum of the vector components.

In short, the lateral load is transferred to the ground anchor point through the tension force in the guys. Remember, the guys are tension force capable only. Depending on the wind direction, only two guys will be active at any one time.

In most cases, there is some moment at the base of the guyed tower, but it's small when compared to the freestanding tower. There is also less horizontal shear, since most of the lateral load is taken to the ground through the guys. The axial load is larger than that of a freestanding tower, due to the vertical vector components developed in the guys resisting the lateral load.

Think of it this way:

■ Freestanding tower — only one load path
■ Guyed tower — two to three load paths

Multiple load paths translates to smaller components or larger load (antenna area) capacity — all other things being equal.

What About Used Towers?

Having presented some tubular legged towers specifications, it's a good time to introduce a concept familiar to hams everywhere — that of buying or utilizing *used* towers. The obvious question is, of course, does this tower represent a bargain, or not? If you're injured (or worse, killed) by choosing and trying to erect a used tower, then the savings were not, in a word, worth it. How can you determine the value of a used tower?

The first (and simplest) choice is to examine the leg bolts from any Rohn tower. If the bolts are rusted on the *inside*, then I would suspect that the interior of those same legs could include damaging rust, as well. I'd pass. I've seen bolts so rusty that they were less than one half their original diameter. I've also seen 20-year-old towers with legs bolts that appeared "like new" on the inside. It's this situation with tubular legged towers that should be your first concern — the integrity of the steel on the interior of the legs. Make every effort to determine their condition. Typically, the Z-bracing will rust before the tubular legs. The Z-braces receive more wear (from climbing and handling), and being solid rod, seem to acquire less zinc than the legs.

Hams are often concerned about surface rust on the steel of tower legs. It's not usually a problem. Simply painting over it with good, cold galvanizing paint will work well. It is not necessary to spend hours grinding away the original hot dip galvanized coating! Indeed, that's probably a bad idea.

Sometimes, enterprising hams will find a local plant providing re-galvanizing services, often "pickling" or dipping the sections in an acid bath first, before coating them. This can work surprisingly well. Or it can be problematic, especially if the re-galvanizing is a little sloppy. The bolt holes and section joints can become filled with excess galvanizing (sometimes called *slag filled*). Or the coating can be too thick, and sections will no longer mate together. This is a difficult decision, as most plants will require a minimum order — so you cannot try out a few sections beforehand to see how well it works. I've been lucky, with both tubular tower designs and AB-105 style (bolt together) sections, however, in having such work done myself.

If the bolts are not available (which is often the case with tower sections that have been taken apart), then it's a real crapshoot in terms of knowing exactly what's on the inside. Try to pay particular attention to the tower legs, looking closely at the ends. It's not unheard of to shine a small spotlight down the leg, trying to gauge the amount of rust or wear that's taken place. Again, it's your life that's at risk here. Don't be afraid or ashamed to be super-critical!

I've also encountered Rohn 25G towers which, although they'd been up for years, still contained the original, factory-packed, legs bolts tucked inside! This can sometimes be a problem as water can collect, freeze, and then split the tubular leg at this point.

Another more common problem is found with a base section that is buried in concrete. The base burial section can develop a crack because the original installer did not provide a gravel bed under the legs for drainage. If you don't allow for drainage in cold climates, water can collect inside the legs and freeze, expanding and cracking the steel tubing. Depending upon their size (they're usually small), these freeze cracks can be a problem. I've repaired many of them, and the section has gone on to serve me (or a client) well for many more years. The solution is to drill two small holes, one at each end of the crack (to prevent the crack from spreading), and then close up the crack. This involves *careful* work with a hammer and a punch, and then carefully welding the crack closed. Then, a few coats with a cold galvanizing spray paint, and the section can be re-used. I have also thrown sections away — ones with cracks or damage I considered too serious to attempt repair.

Again, it's false economy to bet your life on this stuff, especially when it's the very first section of tower in a stack, reaching upward, toward the heavens — holding your hard-earned investment in steel and aluminum!

More information on inspections and maintenance and working with used tower is presented in later chapters.

In the Air:
The Realities of Climbing

Although most states regulate professions that involve health and safety (barber, electrician and so on), they require no license or certification to climb a tower.

And while there are such things as OSHA (Occupational Safety Health Administration) guidelines for commercial climbers and workers, for the radio amateur there exist only the good intentions and well wishes of his or her friends and relatives. That's risk, personified, indeed!

What can you do? How can you know or ensure or even determine you're doing "the right thing" when it comes to doing work on your own tower — whether it's erecting the tower or making repairs or stacking a variety of antennas on it? Who are you going to call?

This chapter will attempt to provide you with not only the answers to these questions, but some guidelines — a process or set of procedures that will allow you to do such work not only safely, but with confidence.

With experience, you'll be as comfortable atop the tower as the climber in **Figure 3-1**. First, though, we'll go over some of the gear that will help you work safely and comfortably on the tower.

Figure 3-1 — Proper gear, physical conditioning and experience are key ingredients in working safely and comfortably at height. This chapter explores what it takes to be successful. (K1SFA photo)

Dress for Success

Clothing choices for tower work are not as arbitrary as you might think. Safety considerations should always drive your decision-making. Except in the very hottest weather, long pants and long sleeve shirts make sense and they should be made of heavy material such as denim. I've spent long days working in the high heat of summer in South Texas, where even shorts and tee shirts seemed oppressive. Upon reaching the ground, I've found enough nicks, scrapes and scars to warrant such heavier clothing. Think of appropriate clothing as an investment.

Let's begin, keeping in mind *'tis better to feel good than to look good*, contradicting Fernando's comedic suggestion otherwise. (For those who are not Saturday Night Live fans, Billy Crystal played the hilarious and outlandish character Fernando, inspired by the real-life actor Fernando Lamas.

Fernando was known for saying, "'Tis better to look good than to feel good.")

Footwear

Although I've seen friends climb wearing simple flip-flops, zipping right up Rohn 25's tiny Z-bracing (yes!), the simple fact of the matter is that you need not only adequate protection for your feet, but some support as well. A steel shank insole is a very good feature if you'll be spending any extended time at all aloft.

I prefer what the industry terms *linemen's high-top boots* such as the boot shown in **Figure 3-2**. Their heavy leather construction withstands the rigors of climbing, and their heavy steel shanks (originally designed for the bolt steps of utility poles) provide plenty of support. I bought a pair years ago to be compatible with my climbing gaffs (spikes used to climb wooden poles or trees), and by now, they're second nature with me. Even these rugged boots will do little to mitigate against the wear, tear and extreme awkwardness of X-braced towers — Rohn BX-series, the old Trylon AB-nomenclature towers installed without climbing steps, and similar designs.

A few years ago, I invested in some Carolina Boots brand professional pole-climbing boots. My first pair I found at Sportsman's Paradise, a surplus house, for $90. That seemed costly at the time. When my entire job-box was stolen from the bed of my pickup, I lost them and had to buy directly from the factory. At $200 for the new boots, the first pair was an obvious bargain. They are not too hot for

Figure 3-2 — The lineman's boot offers good support and protection for feet and ankles.

summertime wear; they're comfortable.

I've also invested in some high dollar socks with moisture-wicking abilities and padded soles. Again, I've not regretted it. I wear sock liners underneath high top (over the calf) socks. I roll the socks down over the loops of the boot laces, helping to ensure they don't snag on something while I'm working. Again, these are worth every penny.

A Clothing Ensemble

My approach to clothing has always followed the ensemble idea (no laughing now, from those who know me outside of tower work). The ensemble idea incorporates clothing and equipment items that offer protection while still providing for functionality. Here's a quick overview, followed by some detailed suggestions.

Protection is just that, and should cover you from head to toe: hard hat, shirt, jacket, gloves, pants and appropriate footwear. Weather and seasonal conditions will, of course, dictate some of your choices.

We've mentioned long pants, specifically denim or jeans. I like the Carhartt brand work pants, as well as typical blue jeans. I've always been a big Carhartt clothing fan, ever since getting my first "chore" coat as an Ohio farm boy. I remember being pretty disappointed when I outgrew the coat — it was that comfortable and functional. I truly like their 12 ounce canvas material pants. The ad copy is right on the money: "The relaxed room legs and large, gusseted crotch of the pants offer unparalleled freedom of movement…"

Where I live, summers can be brutally hot, and while I don't like the idea, shorts sometimes make good sense. Carhartt makes shorts with the same material, but without the long legs, meaning I've got shorts I seemingly cannot wear out. I wear them a lot in the hot days of July, August and September, even while not on a tower. Another good choice is cargo style

shorts from Cabelas sporting goods stores. They have angled, slash-style pockets, making inserting and retrieving things especially easy. (I remove the hook-and-loop fasteners to make getting in those rear pockets as easy as possible.)

Gloves

Gloves remain a sore subject with me. I started wearing gloves several years ago — on every climbing job, despite the weather. (Mostly, I got tired of the cuts and scrapes.) And, while I continue to seek out the "best" glove for this type of work, I'm currently partial to the mechanic's style — with the index, middle finger and thumb tips removed, providing the needed level of dexterity when dealing with small screws and the like. **Figure 3-3** shows a well used pair.

Prices typically range around $25 per pair. I've tried them all and found few winners. The Ironclad brand is fairly good, but every pair seems different. I've worn some for months; others have come apart after a couple of jobs. The Mechanix brand Framer model gloves are pretty good, if you can get past the gigantic lettering emblazoned on the back of your hands. They're the coolest in the summer and have not ripped at the seams, but the fingertips have disintegrated over time. Some brands have padded palms, which are helpful, especially as you're playing out a long rope through your hand. Duluth Trading Company often

Figure 3-3 — Mechanic's gloves with the index finger, middle finger and thumb tips removed protect the hands yet allow manipulation of small hardware. These are the author's gloves after just two months of use.

has good information and good deals on not only gloves, but also other construction wear-suited clothing.

I sometimes get asked about "waterproof" gloves and sometimes suggest Playtex as the answer. This attempt at wit often fails to work, but folks *are* seemingly serious in asking this — thinking there is such a thing. But guys, there's a big *hole* in one end of the glove, so no amount of space-age material or computer aided design is going to make any glove waterproof. Yes, if you're going to dip your hands in the sink there are options, but working in the real world means you're going to get your hands wet. It's just that simple. I know, we're not supposed to be working in the rain or other compromising conditions, but sometimes circumstances (and demanding clients) dictate that we work in inclement weather.

Undergarments

Perspiration can leave you chilled, no matter how well your outer shell fends off falling moisture. Of course, such language is part of the basic concept of layering clothing. Since cotton does a good job of retaining perspiration, it makes sense to wear underwear ("base layer") that's made from some other material, to transport the moisture away from your skin. You stay warmer even as you sweat.

Fortunately there is a wide variety of underwear designed for people engaged in all forms of outdoor activity. Garments made from silk, wool and synthetic fabrics are available in various weights and shapes. REI's MTS, or Patagonia Capilene, or Polartec Power Dry are some suitable brand names. Check stores or Web sites that cater to outdoor or winter sports enthusiasts.

In putting together your layering system, remember that what you are trying to do is create a system that effectively allows for breathability (wicking away sweat), rapid drying, insulation, durability, wind-resistance and water-repellence, while still being lightweight and allowing freedom of movement. What you choose on any particular day will depend on your intended workload and Mother Nature. Choose wisely and you'll fool her every time!

Hard Hats

The most overlooked area of the body is, of course, the one that's most easily damaged — your head. Hard hats (**Figure 3-4**) are a fact of life around construction sites, yet I rarely (if ever) encounter hams who own them, let alone use or wear them. They're not that expensive, and well worth the effort to acquire and use. Here's how they work and why you should wear one.

A conventional hard hat consists of two components, the shell and the suspension, that work together as a system. Thermoplastics (polyethylene, polycarbonate) and thermoset materials (fiberglass and phenolic-impregnated textiles) are commonly used as shells of industrial hard hats. These materials have proven to be durable, reliable and lightweight, while still providing effective protection. The shell should be inspected routinely for dents, cracks, nicks, gouges or damage from impact, penetration, abrasions, rough treatment and wear that might reduce the degree of protection originally provided. Any hard hat showing signs of worn or damaged parts should be removed from service and replaced.

All hard hats are susceptible to ultraviolet light damage, temperature extremes and chemical degradation. Thus, users who work in excessive exposure to sunlight, heat, cold or chemicals, should replace their hard hats more frequently. Degradation of thermoplastic materials may be apparent when the shell becomes stiff, brittle, faded and dull in color or exhibits a chalky appearance. A hard hat should be replaced immediately at the first sign of any such condition. A simple field test can be performed to determine possible degradation of polyethylene shells. Compress the shell inward from the sides about an inch with both hands, and then release the pressure without dropping the shell.

The shell should quickly return to its normal, original shape, exhibiting good elasticity. Compare (if you can) that elasticity with a new shell. If your hard hat does not match, replace it at once.

The hard hat's suspension system (**Figure 3-5**) is just as important as the shell. Its main purpose is to help absorb the shock

Figure 3-4 — Often ignored or forgotten the simple hard hat is an essential piece of safety gear.

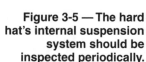

Figure 3-5 — The hard hat's internal suspension system should be inspected periodically.

from a blow. Therefore, it must be in good condition at all times. Like the shell, the suspension must also be inspected and replaced periodically. Over time, the suspension will become worn and may be damaged. The suspension should also be inspected for cracks, frayed or cut straps, torn headbands or size-adjustment slots, loss of pliability or other signs of wear. Perspiration and scalp oils both help contribute to this sort of wear. Any suspension that's damaged should be removed from service and replaced immediately.

It's impossible to predict a specific time frame for hard hat replacement. Some large corporations replace their hard hats every five years, regardless of outward appearance. Where environments are known to include higher exposure to temperature extremes, sunlight or chemicals, hard hats should be replaced after two or three years of use.

We've all heard or read that heat loss happens mostly through our head. Debate continues about the amount of heat lost that way, but it's undeniable that about 15% of the body's blood volume is in the head, and the brain (which regulates *everything* our body does) needs not only the right amount of blood, but blood at the correct temperature. So, the *Merck Manual* tells us, 30% of our body temperature can be lost through the head and neck. For the tower worker, who's often wearing a hardhat, this can be a problem. But it's an issue worth resolving — typically, through an appropriate liner, which can prevent heat loss and also aid with cooling in the summer.

In the winter, I've taken to wearing a simple balaclava under my hard hat — a black nylon hood that some ham pals (jokingly?) refer to as my attempt to appear Ninja-like. But the simple cloth facial covering truly works and keeps me warm well into freezing temperatures.

The Climbing Harness

As with many things, tower climbing times have changed, and the materials and tools used in those duties have changed, as well. Today, the synthetic-

webbing *harness* rules. The days of the old, simple lineman's belt (**Figure 3-6**) are over. As you can see from **Figure 3-7**, the modern harness is quite a bit more involved than a simple belt.

Along with these changes, confusion seems rampant, especially among hams, who have heard or read or otherwise received some information or partial details concerning rules or "certification" or some other regulations. Let's try to clear up the confusion.

First of all, there are no OSHA certifications for anything. What OSHA does have are *training requirements*, which are included or embedded in many of their standards. It's an employer's responsibility to train employees where required by OSHA. If the employer provides such safety training, it is documented and then these trainees are considered "certified." OSHA does not actually certify anyone. The employees simply receive a card saying they've received such training.

For tower climbers, employers must have a written fall protection plan. Employers must train their employees who will work at height and use such fall protection. OSHA requires the use of a full body harness, not just a simple belt.

Here's the actual regulation wording:

1) The employer shall ensure that each employee under the fall protection plan has been trained as a qualified climber.
2) The fall protection plan shall be made available and communicated to exposed employee(s) prior to the employee(s) beginning work, and such communication shall be documented.
3) The fall protection plan shall identify each location on the tower structure where fall protection methods as described cannot be used. As soon as adequate tie-off anchorage points or other fall protection systems can be established, the employer shall utilize any of the fall protection systems.

Figure 3-6 — How *not* to climb a tower. I found this on the Internet once upon a time. After reading this chapter, you'll be able to spot the problems. For starters, the old lineman's belt should no longer be used. Instead, climbers should be wearing a personal fall arrest system (PFAS) harness. What else should this climber do differently?

Figure 3-7 — The DBI/Sala Exofit XP harness is designed for tower climbing, with leg loops, a chest strap and shoulder straps in addition to a waist belt. The D-rings at the center (the part that looks like an old lineman's belt) are for a positioning lanyard. The fall arrest lanyard connects to a D-ring in the center of the back (see Figure 3-8). The outer D-rings connect the positioning lanyard to a seat, allowing the climber to relax a bit during extended sessions.

The critical thing you should take away from reading the regulation is that this rather simple-sounding *harness* is, in fact, part of a system — an entire set of tools or gear designed to keep you safe. It's referred to as a *Personal Fall Arrest System* (*PFAS*).

Elements of a PFAS System

PFAS gear is a critical component for safe and productive tower work. You inspect that gear before climbing, every time (following the manufacturer's supplied guidelines)! There are three critical components for any PFAS:

- The anchor points you select on the tower.
- Body harness — properly fitted and inspected.
- Connecting devices — properly selected and inspected.

Typically, harnesses will still allow hooking in a lanyard to *D rings* at the hips (holding the worker in place on a tower). There's also another D ring in the middle of the back, where a second lanyard connects — a shock absorbing *fall arrest* lanyard. The lanyard at the hips is only a "positioning device." *It is not fall protection!*

Figure 3-8 shows a close-up of the back D-ring with two fall arrest lanyards attached. The other end of these lanyards are connected to locking carabiners (**Figure 3-9**). The locking carabiners can be attached quickly to tower anchor points. **Figure 3-10** shows the PFAS system in use, protecting a climber on a tower so he can use both hands to work.

All PFAS harnesses are designed to save your life, should the unthinkable happen while at height. This is where and how times (and thinking) have changed — no one climbs a tower intending to have an accident. But accidents can and do happen. We all know that. Fortunately, for today's climbers, should that unthinkable thing happen, they can be protected.

Choosing a harness will primarily mean picking one that's comfortable. And the best way to do that is,

Figure 3-8 — The heart of the PFAS system is the D-ring in the back of the harness for connection of a fall arrest lanyard. This version has two safety lanyards that can be attached to different anchor points on the tower.

Figure 3-9 — The other end of the fall arrest lanyard has a locking carabiner for quick and sure attachment to the tower.

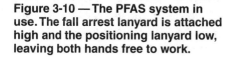

Figure 3-10 — The PFAS system in use. The fall arrest lanyard is attached high and the positioning lanyard low, leaving both hands free to work.

of course, by actually trying the harness on. Unless you live near a large metropolitan area, that's probably impossible. You're looking for comfort in the shoulders, upper thigh and seat areas, adequate adjustability with the leg straps, and some good padding in the shoulders, the belt and the saddle. Other things to consider (which will quickly become essential, once you actually start using your harness) are:

- How many tools pouches can you attach to the belt?
- Do you want a built-in tool belt or floating tool belt?
- Do you want a twin-leg lanyard?

This is a good place to caution you not to be taken in or swayed by gear that claims to be "OSHA Certified." Because, once again, OSHA doesn't certify any equipment. You won't go wrong buying your climbing safety equipment from manufacturers such as DBI/Sala, Miller, MSA and so forth. Be prepared to pay for their quality and reputation. But it's cheap insurance, plain and simple.

About that Old Lineman's Belt

You may wonder about that shift from the old lineman's belt. The reasons are numerous and overwhelmingly simple. Studies have shown (proven, in fact) that the simple, single belt, worn securely at the waist, can cause serious problems and damage to a climber should an accident occur and that climber falls.

If the climber does not fall *off* the tower, but is secured to it (by the typical positioning lanyard), the belt can slip up, onto the chest area. That can constrict the diaphragm, making breathing difficult or even impossible. Plus, the physics of falling "into" that waist-level secured belt can harm other parts of your body. Damage to your internal organs is likely.

Of course, no one ever expects or wants to fall from a tower. Of course, no one plans an accident, either. It's just this simple: the shift in thinking and these applications toward a more safety conscious workplace make very good sense.

What works for industry professionals can work for us hams, too. The idea of "100% connected" or always being tied off to the tower is a good one. But with Rohn 25G and 45G, the tie-off point must include two welds and the side rail (the tower leg) in order to be secured to that required 5000-pound structurally-rated member — something that's not only not necessarily easy, but sometimes impractical, as well. Simply clipping on to the Z-bracing is *not* an answer — it's simply not strong enough!

Climbing — Safety is Always the Primary Concern!

Once you're ready to climb, and you reach out, grasp the tower and step up, it's time to remove all other thoughts, worries or concerns from your mind. Your concentration on climbing should be complete and total. You should be focused on what's right there in front of you: the steel, the placement of your hands and feet on it, and your safety gear and its attachment to that tower.

It makes sense to climb a short distance, stop and get acclimated to your surroundings. It may seem overwhelmingly obvious, but being off the ground (even a few feet) is a very different environment from simply walking around or standing erect. Your body and mind both need some time to adjust to it. The higher you go, the more this is true, of course. Paying attention is critical!

There's really no "boss" or supervisor to tell you what to do while climbing. It

Figure 3-11 — Climbing around obstructions is safer using two lanyards. Once the fall arrest lanyard is connected above the guy wires, the positioning lanyard can be unhooked and the climber can move up and over the guy attachment.

is, in the best sense of the word, a liberating experience. Climbers will be called upon to solve problems and do things in an environment where the average person will never go — that freedom can become an addicting way of life. Doing such work successfully is strong motivation, along with a very positive form of self-expression, yet that's not something most climbers would admit to or even perhaps consider.

Climbing is a strenuous physical activity. It requires practice, some serious skills and paying attention. You climb with your legs; you don't climb with your arms. That's not to say that upper body strength isn't good to have (especially if or when you're pulling yourself up and around star guys or ring rotators), but the sheer work of climbing up and down is done by having your legs drive your hands.

As you climb, your body should be relaxed and limber. There are no "rules" regarding how to grip or grasp the tower (overhand, underhand or a combination of the two). The critical factor is to do what works for you. Develop something, call it a system or whatever, and then stick with it, climbing that way each and every time. If this sounds like I'm suggesting you actually practice climbing your tower, you're right. You should be familiar with where you'll grasp what on your own tower — intimately so — before you decide to go up there to perform some work. Practice, of course, is also strenuous, but it pays dividends in the long run.

The old mantra "train like you work, work like you train" just makes very good sense. Climbing needs to become almost second nature if you are going to succeed doing it.

Making the Climb

As you climb, always look to confirm that your safety hook is closed. Never rely on just the sound the latches make, regardless of what you may have been told or read someplace. Always check visually! **Figure 3-11** shows how to use two lanyards to get safely around an obstruction and **Figure 3-12** shows how to make sure you're attached to the tower while you're climbing.

Always think about your next move, before you make it. Always take your time. Climbing isn't a race, and what works for one climber may seem horribly fast (or slow) for another. Think of it as your commute to work, which will still be there once you arrive. The idea is to get there relaxed and ready to work, and not be so tired you cannot do what you climbed up there for in the first place!

As you work your way upward, you should be keeping a similar spacing between your body and the tower itself. About nine to 12 inches from your belt buckle is a good working distance. Again, as you climb, your legs drive your hands — your hands should be the last things you move as you ascend from step to step. (When descending, working your way down, your hands lead your feet — they move first. And, when descending, you simply let your legs and feet "fall" in to place; don't

Figure 3-12 — Proper use of the safety hooks attached to the D-ring in the back of the harness. Attach and climb, attach and climb as you make your way up the tower.

push them downward.)

If you climb your tower (or similar towers) often enough, you will develop what's often called a sense or muscle memory. Your body will recognize where things should be and how they should feel. This does not mean you shouldn't or won't be watching what you're doing or paying attention. It's merely one more method of knowing where you are at all times.

Once you reach the top or wherever you will be working, it's wise to stop for a moment and rest or catch your breath. This is an ideal opportunity to look around, to scope out where you'll be working, to determine if you can do exactly whatever work you pre-visualized on the ground.

Any talk about climbing would be incomplete if it did not address the following: Once up there, you *must* be able to let go, to release your grip on the tower and have both hands free to do work. I've climbed with guys who couldn't do that and we both found it frustrating, trying to accomplish even the simplest tasks with only three hands available. You *must* have confidence in your harness — not only to protect you from falling, but to secure and position your body perfectly, so that you can move around, work and fully function up there.

This is also an appropriate place to say there's nothing wrong with *not* being able to do any of that. Not everyone is cut out or equipped or able to climb and work at heights. (Plenty of people say I must be nuts to be doing such work as I approach retirement age!) But, if you wish to do such work successfully, you must be able to put your trust in your gear, in your own physical and mental abilities and condition, and your knowledge and perception of that vast space you find yourself in. You must be able to simply let go and do it. As I say, it's a liberating experience.

Ground Crew Considerations

Those who remain on the ground are essential parts of any successful tower or antenna project. What makes for a

good ground crew member? The ability to assess and analyze situations quickly and the ability to anticipate what the tower climber(s) will want or need. Crew members also need to be adept at working with a variety of ropes, tools and rigging situations based on those needs. (A "retired" climber makes an excellent ground crew person!)

Ground crew personnel need to realize their work will be filled with long periods of interminable boredom — just standing around waiting. Waiting for the climber(s) to do something, to finish something (which will then require action from *them*), to want something or to need something. When that happens, they must work quickly and correctly. (For the ground crew, standing around talking and doing nothing is fine, up to a point. But when the crew aloft wants or needs something, their perception will be that standing around is *all* the ground crew does if they're not receptive to requests and not quick responders.)

During the pre-climb "tailgate meeting" it's important to make sure everyone's on the same page in regard to the work that needs to be done and how it will be accomplished. It's especially important to talk about the verbal commands used by the crew on the tower. And by the way, those who are on the tower are always in charge — they are closest to the action, can see better than anyone what's going on and it's their lives at risk. *They call the shots. Always.*

Here are the voice commands I use:
- *Slack* means to loosen the line.
- *Tension* means to tighten the line. (Once you have tension on the line, simple up and down commands take over).
- *On the tower* means the load has been safely secured to the tower (ground crew can let go of the rope).
- *Standby* means hold the load or line — to just keep doing what they're doing when they hear the command.
- *Heads up* or *headache* means you've dropped something.

Of course, the real goal is never to

drop anything, but such things do happen from time to time. Proper rigging and experienced personnel will help ensure that the ground crew folks are not standing next to or near the tower's base anyway.

Communication

Communication between climber(s) and ground crew is the name of the game, of course. That can sometimes become problematic, especially when the tower crew is working up high — above 100 feet. Wind noise rises quickly above the average tree line. Vehicular traffic can be a problem in some urban environments, as can airplanes (which usually go away quickly, but Murphy often plans for their arrival, or so it seems). Dogs, chainsaws and lawn mowers and other assorted household noisemakers can be troublesome or annoying.

I've tried all the typical solutions including FRS radios, 2 meter handhelds, those little VOX-headset 47 MHz radios. They all work, to some extent, but they all have shortcomings. Shouting, however, isn't the best answer — but it's often used, especially as a last resort or in moments of extreme exasperation or in final frustration. Whatever you do, try to come up with something that works for you and your crew.

Crew Size

Ground crew size will be determined by the scope of work. Larger jobs, such as maneuvering the large beam in **Figure 3-13**, require more hands. Sometimes, tower work can become dangerously close to a social event, with all the ground crew folks standing around, chatting away or sharing stories. While this is not necessarily bad, it can be distracting to the tower crew. It can even hinder progress if the tower crew has to wait for the ground crew to catch up with their needs.

It's important that all workers pay attention to water and food needs. Dehydration can occur quicker than you might think, especially in hot and dry summer months. Make certain you provide plenty of drinks and fuel for everyone. Water is better than soda and even the high-electrolyte sport drinks. By all means, save the beer for when the job is finished, never during or before the work itself!

Whatever you do, do not climb or work alone! Not only is it unwise from a safety standpoint (someone needs to call for help if you are injured), but also it's almost certain that at some point you will need that second pair of hands at ground level. Ground crews sure beat climbing down and back up!

And finally, a few words on the idea of tower/antenna parties at a fellow club member's house over the weekend. This oft-used phrase (and activity) gives me pause whenever I hear it mentioned on some 2 meter repeater or at a club meeting. The importance of good help when doing tower jobs cannot be taken for granted. Once again, tower work is dangerous, and your life depends on those ground assistants. It's a Catch-22, for everyone has to start somewhere, but take care when choosing your helpers. Surprises are *not* good in such work sessions, and inexperienced help can create some dandies. Again, the watchword is safety. Experience can go a long way toward making sure everyone, on the ground and in the air, remains safe.

Safety Wrap-up

This chapter is intended to focus your attention on safety, which is so often overlooked. OSHA has no jurisdiction over your ham tower, and they won't investigate you should an accident occur. No government official will come by to check on you, to see if you are qualified to climb your tower, or if a climber you've hired is, either. It's up to you to learn about safe tower work practices, to equip yourself with the right clothing and tools, and to pay attention and perform the climb carefully.

Regarding insurance, if you have someone work on your tower, you are not required to make sure they're trained. As with hiring any contractor, it does make sense to ensure they have liability coverage. It's okay to request a copy of their certificate of insurance — to prove they have such coverage. Again, this all sounds like so much common sense in many ways…and in many ways it is.

Yet, the simple fact is that these safety concerns are all-too-often overlooked. Within industry, employers and workers both seem to share the misconception that simply having the right safety gear and fall protection hardware is enough to keep everyone safe. Training falls by the wayside all too often. For the occasional tower climber — the typical ham radio tower owner certainly applies here — training may easily be forgone, or more likely, simply forgotten.

Working at height is dangerous — so much so that safety should always be your main concern! Safety should drive every action, both up on the tower, and among those workers aiding you on the ground. The idea is to accomplish the work, and then be able to enjoy the fruits of that labor, back on the ground, in the shack, and on the air.

Safety — thinking about it, practicing it, and reviewing it — makes that possible.

Figure 3-13 — The ground crew maneuvers a large beam into place at W9GE. The antenna's many elements, sloping terrain and nearby trees require careful attention.

CHAPTER 4

Tower Bases and Guy Anchors

Once you have your tower plan finalized, it's time to start digging dirt and pouring concrete — the subjects of this chapter.

Base Types

The base for your ham radio tower is literally its foundation — what supports and carries the weight of the tower itself, along with forces directed down the legs from the winds. Bases vary, according to the type of tower. Four commonly used guyed tower base designs are shown in **Figure 4-1**: the tapered tower base, the pivoted tower base, the composite base and the pier pin base.

Hams often use a variation of the composite base, where they bury a section or part of a section. A special "base burial" section is simply a smaller-than-normal section sized to bury in concrete. Hams also commonly use the pier pin base. A pin in the center of the concrete pier protrudes through a hole in the center of the base plate

and holds it in place. The base plate is not bolted down or anchored to the concrete, though, allowing some side-to-side movement. The tapered tower base is seen on larger ham towers such as Rohn 65. Pivoted bases aren't too common in ham installations.

Freestanding towers utilize heavier construction than guyed towers and taper in toward the top from a wide, spread apart base section. Freestanding towers exert larger weight-bearing forces on their foundations than the

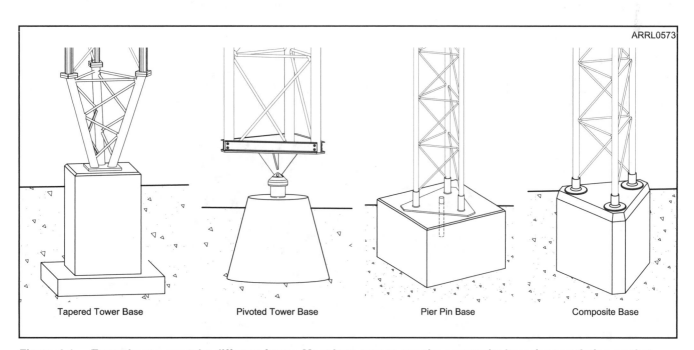

ARRL0573

Tapered Tower Base   Pivoted Tower Base   Pier Pin Base   Composite Base

Figure 4-1 — Tower bases can take different forms. Most ham towers use the composite base (or a variation on that with a buried base section) or the pier pin base.

typical guyed tower. So, larger and deeper bases are required. Some base designs have each leg supported by an individual foundation, for example.

Buried Section vs Pier Pin

The most commonly used base for guyed towers is probably the buried section. A lot of hams don't like the idea of burying a perfectly good, working section of tower in a chunk of concrete. I can appreciate that. But the buried section will simply feel more solid than any other method, and it's oh-so-easy to install.

Pier pin bases, whether they use a tapered tower section or a square base plate and standard section, require further attention to detail. The concrete for the base plate (if used) must be perfectly level. The pin must be the proper one — designed for this application. You don't simply insert a bolt from the local hardware store in the concrete and consider it ready. Any tower with a pier pin base will require temporarily guying the sections as you erect them, as there's no support from the base. This can sometimes prove troublesome, certainly time-consuming. (There are no shortcuts here — do *not* think you can simply use ropes to temporarily guy the tower in place as you work your way upward. Only standard steel guy wire should be used.)

Safety Tips

Regardless of your tower type, it's important to keep safety in mind when digging the hole for your tower base. Do not take such work for granted. Each year people die or are injured from being trapped in a hole when soil collapses. The following simple overview of soil mechanics may help you understand some things to watch out for when digging the typical tower base hole.

Soil is a generously vague term. It can mean weathered rock and humus, which we usually refer to as clays, silts and loams. But soil can also include gravel, sand and other rock. Soil can be extremely heavy, weighing more than 100 pounds per cubic foot. One cubic yard of soil (3 feet × 3 feet × 3 feet, or 27 cubic feet) can weigh more than 2700 pounds. If that soil is wet or contains rocks, it will be even heavier. That's like having a car fall on top of you when you're down in that tower base hole!

Keeping that 100-pound per cubic foot weight in mind, we realize that not only does the soil exert this downward pressure, it's exerting some pressure sideways, or horizontally, as well. That pressure is one half the downward force. So, if you're down in the ground five feet, standing next to a wall, there's 250 pounds of pressure coming "at you." Unless you do this professionally, you're no doubt unaware of the pressure and probably have not shored up or otherwise attempted to secure the sides of the hole. Most of the time, that's not a problem. But digging tower holes can be a potentially dangerous job, and you should be prepared.

In undisturbed soil, equilibrium is maintained by having all the horizontal forces push against the vertical forces. In a hole, or trench, that equilibrium is lost because the horizontal forces are removed. The soil may not be able to support its own weight, and may collapse. The first failure will be at the bottom, then slowly (over time) work its way upward. The typical "bell cutting" we often do at the bottom of a tower's base hole (making the bottom wider than the top in a bell shape) only compounds this potentially dangerous condition. Shoring up the walls or sides of an unusually large hole can be a useful safety technique.

Digging holes that are larger than you need, then building a form inside, and then back filling, is sometimes necessary — especially in very sandy soil. It's truly time-consuming, and requires considerable attention to detail. Don't be afraid to ask for advice if you're not certain what type soil you have or are digging in, as this can be critical. Your county extension agent can usually help answer such questions.

A few simple rules: Any time you find yourself further than four feet down in the ground, a suitable ladder or set of steps should be used to provide an exit. The ladder should extend at least six inches above the surrounding excavation grade. While working in the hole, bracing or shoring is a good idea. Of course, the installation of the tower base section and steel reinforcing cages in the hole often prevents this or makes such work highly impractical. And, it almost goes without saying — never work in such conditions alone.

Digging the Holes

Your tower manufacturer will have specifications and drawings for the required tower base. We're talking about a pretty big hole — typically 4 feet deep and 3 to 4 feet square for a small Rohn 25 guyed tower. Base holes go up in size from there; 5 or 6 feet square and 5 feet deep is not unusual for a freestanding fixed or crank-up tower. Before digging the first shovel full of dirt, lay out the location and dimensions of the base hole (**Figure 4-2**). Double-check the location and orientation of guy anchors. The simple, homebrew tower layout tool shown in **Figure 4-3** can help with this task.

Digging almost any tower base hole can benefit from

Figure 4-2 — Paint and string are used to mark the layout of a large crank-up tower base prior to the backhoe's arrival.

mechanical help such as a backhoe, excavator, tractor with backhoe attachment or other earth moving equipment (**Figure 4-4**). With rocky soil, ledge or similar challenges it may be your only choice. If the site is accessible and the right equipment is available, hiring equipment to dig the holes may be less expensive than you expect. Call around, have the plans ready and be prepared to explain exactly what you need. Show the operator the location of septic systems, underground pipes or electrical lines and other obstacles. Depending on where you live, you may need to notify the local utility company or have a service such as Dig Safe mark locations to avoid.

Equipment operators usually charge by the hour, and expect a charge for transporting the equipment to the job site (called "portal to portal"). Machines come in a wide variety of sizes and capabilities — one that can move a boulder the size of a car isn't a good match for digging your 3 to 4-foot-square tower base. Machines may have rubber tracks, cleated metal tracks or tires. The tracked machines can go more places, but those with cleated tracks can quickly make a mess of your lawn or blacktop driveway. Be sure to discuss all of your expectations with the operator and understand any potential collateral damage before starting the job.

The critical factor, always, is the size and depth of the hole. A smaller bucket is almost always preferred, for it allows for more precision. This precision comes at the expense of longer digging

Figure 4-3 — This simple tool is used to plan and lay out the base and guy anchor positions for guyed towers.

Figure 4-4 — Sometimes a large hole is required and the only logical choice is mechanized help. AB7E used this backhoe to dig the base of his self-supporting AN Wireless tower. (AB7E photo)

times. If you're renting equipment or paying an operator by the hour, time can sometimes be a factor, but I almost always opt for the tighter control. Typical tower holes are not that big a job compared to, say, a house foundation or septic system. It won't take a skilled operator *that* much longer to make a precise hole for you.

The best machine operators I've ever found (and used) have been gravediggers, literally! They can dig you a very precisely squared-up hole, without the bottom being bell-shaped. But typically you should plan on "finishing" almost any machine-made hole by hand.

Traditionally, digging in the dirt has been considered a pejorative term —

that if you fail at everything else, you'll end up digging ditches. By hand. Slowly, painfully and mindlessly. Punishing hard labor. So it sometimes seems with tower bases and guy anchor holes in the ground. You may have to do it the hard way if site access is too restrictive for powered tools, for instance. Or perhaps you cannot locate or afford to hire equipment and/or an experienced operator. Something — whatever it is — means those holes must be dug by hand.

Using Hand Tools

For larger holes, such as a tower base, the traditional shovel and pickaxe can be used in combination. Many folks are surprised to learn that various versions of these most basic dirt-moving tools exist. Here are some suggestions to increase the speed, ease the pain, and make digging by hand far less frustrating.

Making a Square Hole

The trenching shovel or spade is very similar to the round point shovel (which most of us know), only it's very narrow — typically only four inches wide. It's used to remove loose dirt from the bottom of trenches dug by a machine or for digging holes and trenches in soft soil. Get one with a long handle and you can easily expand the sides of holes and so on.

While on the subject of expanding holes, let me suggest a clever solution to "squaring up" dirt. (A mass of square concrete is more resistant to movement than rounded-off lumps poured in the ground. Yet the tradi-

Figure 4-5 — The best post hole digger is the auger version, used here to dig a base for a vertical antenna.

tional round point shovel leaves exactly that — rounded-off cuts.) A simple six-inch steel floor scraper will work wonders (especially in Carolina clay) in squaring up the sides of holes, allowing you to put a precise corner or edge in a tower base hole. I'm pretty sure client Rick Low, N6CY, is still wondering how, working entirely by hand, I was able to make a perfectly square hole in the soil of his Reston, Virginia, backyard (the tower base was inaccessible to any mechanized hardware). The scraper allowed me to create perfectly square corners (I laid a framing square against them as I worked my way around) and perfectly straight edges.

Again, most everyone knows the pickaxe. But consider, instead, the mattock, which is a cross between a pick and a shovel. It has a narrow blade on one side (somewhat like a trenching shovel, only smaller), with a pickaxe point on the other side. You use it like an ax. I use a mattock for digging short trenches and removing rocks. With a mattock blade on one side and a pick on the other, a mattock will dig twice

as fast and twice as easily as a shovel.

Another indispensable tool is the digger/tamper, which is simply a heavy steel bar about five feet long. On the tamper end, the bar has a head (similar to a huge nail head) that's used to compact soil. The digger end has a wide, flattened point like a large chisel. Don't struggle with rocks and debris when the digger/tamper/pry bar will help you to remove them easily.

The Posthole Digger

Now, the true test of the dedicated dirt mover — the main tool in his/her arsenal — the posthole digger. Again, many folks know and have used the standard two-handle model. But not too many know (or have used) an auger model (**Figure 4-5**). The auger is faster (in my opinion) and makes a cleaner hole. It's sometimes difficult to plunge the two-handle model perfectly straight down, repeatedly, especially when you're tired. The real benefit of the auger model is that you can change the depth of your hole by simply changing the length of the handle. For instance, by adding a 3-foot pipe nipple between the auger and the existing handle, you're now digging a 6-foot deep hole by hand. Take it off, and you're back to the standard three-foot deep hole suitable for fence posts and clothesline poles.

The two-handle model has another use in tower work. If the base hole is large enough, you can continue to dig with the shovel, but will find it difficult to remove the dirt. The two-handle digger can be used like giant chopsticks to lift out the dirt your shovel has loosened. Each model has its place in digging holes.

In my Ohio farm boy youth, I simply hated the posthole auger. It meant hours of hard, physical labor, usually during summer vacations, when it was hot, too! I'm amazed to find I've searched out and own one (Seymour makes the best auger, in my opinion). Yet I find myself grinning when I successfully dig what some clients thought was going to be impossible or impractical holes in the ground.

Reinforcing Steel

Once the hole is dug, it's time to think about what goes in it. And no, we're not yet ready for the concrete — we're talking about *rebar*.

Rebar refers to those steel reinforcing bars used in concrete construction. The term dates from the post-war American building boom of the 1950s, although reinforcing steel has been in use since the 1920s. Ridges are rolled into it because it's milled to mechanically anchor in the concrete. Rebar is dark gray when new, but rusty orange and scaly after exposure to the weather. It's made from a variety of recycled materials, and its composition is quite variable, which is the primary reason welding of rebar cages is not allowed by building codes. Wire ties are the preferred method of securing such structures. Your tower manufacturer will specify the type, size, amount and configuration of rebar required in the design of a tower base or guy anchor for your application. **Figures 4-6** and **4-7** show a tower base rebar cage under construction.

Rebar is sized by "the 8ths," meaning increments of ⅛ inch. For example, #8 rebar would be 1 inch in size. The most common sizes for typical tower base applications are #3 (⅜ inch), #4 (⁴⁄₈ or ½-inch), and #5 (⅝ inch). The overall rebar diameter is about ¹⁄₁₆-inch larger than the nominal size because the ridges project slightly. For example, #4 rebar might fit a ½-inch hole in thin material, but it might be a very tight fit.

Rebar is available in various grades, or strengths. Grade 60 rebar is the most common for home use and landscaping. It is usually sold in 20-foot lengths (often cut in half by the big box home stores), but shorter lengths are also available. It can also often be cut at the store. Take advantage of that service if it's available. Rebar can be tough to cut without the right tools.

Reinforcing steel is incorporated into concrete construction because concrete, while very strong in compression, can take very little tension force. The reinforcing steel (rebar)

Figure 4-6 — K4ZA and K4DXA begin assembling the rebar cage for K4KL's 89-foot crank-up tower.

Figure 4-7 — K4DXA puts the finishing touches on K4KL's rebar cage. Wire ties are used to securely connect the rebar pieces together.

Table 4-1
Rebar Specifications

Imperial Bar Size	"Soft" Metric Size	Weight (lb/ft)	Weight (kg/m)	Nominal Diameter (inch)	Nominal Diameter (mm)	Nominal Area (in²)	Nominal Area (mm²)
#3	#10	0.376	0.561	0.375 = 3/8	9.525	0.11	71
#4	#13	0.668	0.996	0.500 = 4/8	12.7	0.20	129
#5	#16	1.043	1.556	0.625 = 5/8	15.875	0.31	200
#6	#19	1.502	2.24	0.750 = 6/8	19.05	0.44	284
#7	#22	2.044	3.049	0.875 = 7/8	22.225	0.60	387
#8	#25	2.670	3.982	1.000 = 8/8	25.4	0.79	509

takes the tension force and lets the concrete take the compression force. This force sharing is how reinforced concrete beams and slabs can take the large loads that they do. The strength of the reinforcing bar is developed by its bond link in the concrete mass. Each size bar has a code-specific length (development length) that's necessary to allow the reinforcing bar to develop its full strength in tension. Hooks at the ends of bars are used to help create this link when the physical size of the concrete mass precludes it being developed by a straight bar. **Table 4-1** shows some key rebar specifications.

Tower base builders may encounter the following terminology, especially if they are dealing with a structural steel company, instead of a home improvement supply store. Understanding the language puts everyone on the same page.

Dowels. These are usually L-shaped, with a 90° bend on one end, or straight lengths of rebar.

Corner bars. These are also L-shaped, but with each side of the L the same length.

Offset bends. These range from a simple Z shape to complex angles and are used in reinforcing concrete walkway steps or other changes in elevation within the footing.

Hairpins. These U-shaped rebars are used to interlock two or more individual *mats* of rebar to give lateral strength to the concrete.

Preparing the Base Hole

With the rebar under control, it's time to prepare the hardware that will attach the bottom tower section to the finished base. Depending on your particular type of tower, this might be a short section of tower buried in the concrete, a pier pin and flat steel plate, anchor bolts or brackets of some type. It's always a good idea to study the tower manufacturer's drawings and specifications and get the right hardware for your application. Obviously it's very important to make sure that the base section is lined up and level so that your installed tower will be truly vertical. In some cases you might install the first above-ground tower section with temporary guys to aid this process. Be sure to secure the base section or hardware so that it does not move as you pour the concrete. **Figures 4-8** through **4-10** show several ways of preparing base hardware for concrete.

The bottom of the hole is covered with a gravel bed to facilitate draining water from the tubular tower legs (if applicable) and help prevent freeze

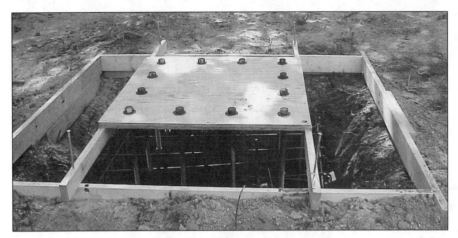

Figure 4-8 — K4KL's crank-up tower base with the forms in place, ready for the concrete trucks. A large sheet of plywood is used to position the tower anchor bolts so that they are hanging down into the concrete the proper distance. The plywood and forms are adjusted so that the base will be level after the pour.

cracks in the steel. Rohn recommends that the gravel be six inches deep for their typical 25G or 45G series bases, but check your tower manufacturer's specifications and follow them.

Although I've seen concrete simply poured into an "empty" hole, it's always better to have a form surround the top of the base hole. This means, of course, it's now time to change from the ditch-digging mentality to that of a carpenter. Remembering that concrete typically weighs about 4000 pounds per yard, your woodworking skills will have to produce a finished form that's sturdy as well as straight. Depending on the type of tower you're erecting, along with the terrain surrounding your base, you'll have some options.

Crank-up tower base forms usually support the erection fixture hardware or simple T-bar bracket over the hole, suspended, if you will, down into the concrete. Guyed tower base forms are simple surrounds if there's a base burial section. If you're using a pier pin base, it's a simple, perfectly flat and level box to hold the concrete.

In either case, two-headed nails will make it faster and easier to take your base form apart once the concrete has cured. Greasing or oiling the insides of the wood will also make for a smoother surface and easier disassembly (I use old motor oil). I use 2 × 6 lumber for the forms, mostly as a convenience, because six inches above grade (above the soil) is the specification Rohn calls for in their catalog. The catalog will

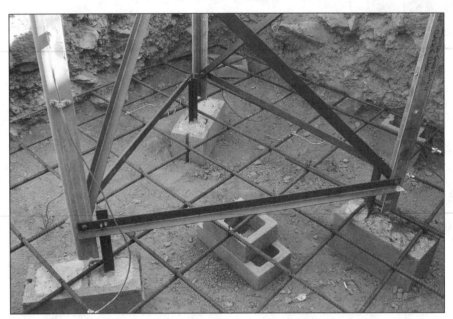

Figure 4-9 — AB7E used these concrete blocks and short pieces of angle iron at the bottom of the hole to level the base of his big self-supporting tower. (AB7E photo)

Figure 4-10 — W7LYQ's tower base is the pad and pier design (a wide pad at the bottom with a narrower pier on top). The buried base section and rebar are secured inside the wooden forms. (W7LYQ photo)

also provide you with the all needed dimensions for the base, of course.

Yes, a little time and attention to detail here will prevent problems from developing later. I'm simply a firm believer that any tower base should be above grade — higher than the surrounding soil. I know the arguments for making them below grade (you may one day sell the house, and so on), but the risks of structural damage and loss of steel from constant contact with dirt and moisture are too great. I've seen too many bases left unattended, sometimes for years, with legs totally rusted through, to ever build anything like this! Base inspection, of course, should be part of any tower's annual inspection process, regardless of whether the top of the tower base is above or below grade.

I like to coat the tower legs and bracing or other steel parts right where they exit the concrete with rubberized auto undercoat spray. I've found that to be an inexpensive and terrific tool to help protect the steel from water and debris that inevitably collects on the base pad.

Now, having done everything, you're ready to pour concrete into that base hole.

Concrete

I've discovered over the past few years that concrete is one of the more misunderstood building materials in the world. The explanations and tall tales I've encountered have been mind-boggling, to say the least. So, let me present a brief treatise on concrete and its application to tower bases. The following sections present some facts, fallacies and tips on working with it properly, whether you buy ready mix concrete or make it yourself.

The Basics

In its simplest form, concrete is a mixture of a paste and certain aggregates — fine aggregate (such as sand) and coarse aggregate (such as gravel or crushed stone). The paste, which is composed of Portland cement and water, coats the surface of the aggre-

gates. Then, through a chemical reaction known as hydration, the paste hardens and gains strength, becoming what we know as concrete. This hydration process is the key, the real "magic" of concrete — it's pliable, plastic and malleable when it's first mixed, but incredibly strong and durable once hardened. This simple, often-overlooked factor is why this material can be used to build anything from your sidewalk or patio, to bridges, dams and skyscrapers, to your tower base.

Having said this, it's probably obvious that the key to success in such a "mix" is the proportions and mixing

process, and that's exactly right. A mixture without enough paste to fill in all the voids between aggregates will be hard to place correctly, and it will be rough and porous when hardened. A mixture with too much paste will move easily and be exceedingly smooth, but will most likely shrink

Typically, the mix should be around 10 to 15 percent Portland cement, 60 to 75 percent aggregate, and 15 to 20 percent water. Trapped air may make up five percent of the mix. Again, it's the Portland cement that's the critical ingredient.

Why's it called that? Because Joseph Aspdin, an English mason, who patented the product in 1824, named it that, after the color of the natural limestone quarried on the Isle of Portland, in the English Channel. Portland cement is manufactured by heating a slurry of limestone or chalk with clay in a kiln, and then grinding the resultant clinker to a fine powder and adding gypsum.

Options for Concrete

Mixing your own concrete with Portland cement and coarse and fine aggregates is perhaps the most economical method, but you must have the separate materials on hand and you must measure them accurately.

Bags of premixed concrete available at home centers are convenient for small projects and available in various strengths from 3000 to 5000 psi. Just mix with water following the directions on the bag and pour in place. Premixed bags will save you the effort of buying and measuring cement and aggregates, but this convenience comes at a price.

For either of these options, a powered mixer is a good idea, and they are usually available for a nominal fee at the local tool rental center. You'll also need water for mixing the concrete (along with the subsequent final cleanup), a source of power for the mixer, and a way to get the mixed materials to the tower site.

Ready mix means just that — the concrete is delivered to you already mixed, per your supplied instructions regarding strength and slump (see the Concrete Terminology sidebar), ready to go into your prepared base and anchor holes. You'll still need to get the concrete into the hole, and that can be a chore if the truck can't get close. Price it out, and you may find that ready mix is reasonably priced compared to buying the ingredients and renting a mixer. Then consider the hard labor involved in mixing it yourself, especially if you need a couple of yards or more. The ready mix plant has a lot more experience in mixing the ingredients, too.

Regardless of what type concrete you choose, the key to success is in the proportions and in the mixing process. As noted previously, a mixture without enough paste to fill in all the voids between aggregates will be hard to place correctly, and will be rough and porous when hardened. A mixture with too much paste will move easily and be exceedingly smooth, but will most likely shrink. Concrete's strength depends on the water/cement ratio. People who don't work with concrete regularly tend to use too much water to make it flow better, which can result in weaker concrete.

Curing Concrete

After your concrete is mixed and placed (poured into your base hole), satisfactory moisture content and temperature should be maintained when the concrete is *curing*. Proper curing guarantees quality concrete, and the curing determines durability, strength, water and abrasion resistance and stability, along with the ability to withstand freezing and thawing and chemicals.

Curing aids the hydration process. Most freshly mixed concrete contains considerably more water than required for hydration, but evaporation (on those hot summer days, for instance, just when you're likely to be putting in a tower base) can

delay hydration. The process is relatively rapid during the first few days, when it's most important to keep the curing concrete moist. (This is why you'll see burlap bags or sheeting or straw strewn across forms, along with periodic spray applications of water, and so forth.) Good curing means you should try to prevent or reduce such evaporation. And yes, curing takes time — the recommended waiting period to reach full strength is 28 days!

By the same token, if it is too cold, the cement will not hydrate fast enough and the water in the mix may freeze. If that happens, the concrete will not set and develop the desired strength. Insulating with plastic sheeting or straw will help to contain heat given off during hydration.

Ordering Concrete for Tower Bases

Here's what the Rohn catalog (which is also their technical manual) says about concrete for tower bases:

1) Concrete materials shall conform to the appropriate state requirements for exposed structural concrete.
2) Proportions of concrete materials shall be suitable for the installation method utilized and shall result in durable concrete for resistance to local anticipated aggressive actions. The durability

Figure 4-11 — Finishing the concrete in W7LYQ's base after the pour. Note the bracing to keep the forms from moving under the weight of the concrete. (W7LYQ photo)

requirements of ACI 318 shall be satisfied based on the conditions expected at the site. As a minimum, concrete shall develop a minimum compressive strength of 3000 PSI in 28 days.
3) Maximum size of aggregate shall not exceed ¾ inch; size suitable for installation utilized; or one-third clear distance behind or between reinforcing.
4) Minimum concrete cover for reinforcement shall be 3 inches unless otherwise noted.

Here's what you need to know, or what you need to do, to comply with those Rohn specifications. First, you'll need to decide if you're going to use ready-mixed concrete — recommended for three yards or more. You'll need to tell the supplier how many yards you require (concrete is measured in cubic yards). You can determine that from:

Cubic feet = Length (ft) × Width (ft) × Depth (ft)

Cubic yards = cubic feet / 27

For example, say your tower base is 4 ft × 4 ft × 6 ft = 96 cubic feet. And thus, 96 / 27 = 3.6 cubic yards required. Be prepared — a yard of concrete weighs around 4000 pounds!

The dispatcher will ask you for the *rating*. "Normal" concrete (sidewalks, patios and so on) is usually rated at 3000 pounds per square inch (PSI). For most tower bases, 4000 to 5000 PSI concrete will suffice, but check to see what your tower manufacturer specifies, and order accordingly.

The dispatcher will ask you what *slump* you require (remember, slump refers to a way of measuring the workability of the concrete mixture). The slump test is done to measure the consistency from truckload to truckload. Good practice is to have around a 4-inch slump for tower base pours.

Figure 4-12 — Once the concrete has set the wooden forms are removed, leaving a beautiful block of concrete. The massive block is needed to support a 70 foot AN Wireless self-supporting tower. (W7LYQ photo)

Figure 4-13 — The finished base after the hole has been backfilled. Note that the top is well above grade to keep dirt and moisture away from the tower legs. Cables will run to the shack via buried conduits. (W7LYQ photo)

Figure 4-14 — Using a crane to set W7LYQ's 70 foot tower on the base. (W7LYQ photo)

The more water that's added to the concrete mixture, the higher the slump, then the lower the concrete's strength and so on. Water reducing agents can be added to make the concrete easier to work without adding water.

Concrete is normally priced by the cubic yard, with the per-yard price based on the assumption of a full load. Note that a typical concrete truck carries 8 to 10 cubic yards, more than you'll need for a typical ham tower base and guy anchors. Be prepared for minimum charges or a "short load" fee if you're only getting a few yards.

Tips for a Successful Pour

You were careful if and/or when inserting or installing any needed rebar cage. You made sure you propped the cage up, off the soil, using suitable spacers. You made sure tubular tower legs (if applicable) were resting on some drainage material, such as pea gravel or other small stone. And while it's sometimes the logical choice (during your summer vacation, for example), concrete should not be poured on extremely hot, dry days, as it will dry out before it can cure properly.

Concrete is a perishable product — it starts to cure as soon as it's made and needs to be poured within about two hours at 60 to 80 °F. You'll need to get your concrete off the truck and into the hole(s) promptly, so plan to be ready to go as soon as the truck arrives. Consider renting a motorized or push-type "Georgia buggy" made for transporting concrete or enlist the help of several friends with heavy-duty wheelbarrows.

Concrete should not be overworked, meaning moving it around excessively with trowels or tools, for example. If it's overworked, too much water will be brought to the surface, which can cause scaling once cured. Try to spread it evenly and quickly once your pour begins. Overfill the forms slightly. I like to use a vibrating tool to work the concrete into corners and so forth, but a 2 × 4 can be pressed into service for this, too. Be prepared for hard work; concrete is heavy!

When working the concrete, I like to leave the center of the base higher than the edges to facilitate water runoff and prevent water and debris from pooling around the legs. Called *crowning*, it's sometimes hard to accomplish with smaller-sized towers, such as Rohn 25, but very worthwhile if you can do it.

A screed board (used to level most forms) cannot be used for the typical tower base — there's usually something in the way, like a burial section, or bolts, or a pier pin. So, you must work with a trowel to finish the base. Concrete finishing provides either a rough or a smooth surface. But your first step is to use a pointing trowel to separate the edge of the concrete from your form. Then, use an edger all along the top edge of the form. The rounded edge left by this tool won't chip off as you remove your form. Hold the tool flat, keeping the front edge tilted up when moving forward, or the rear edge tilted up if moving backward.

Keeping the concrete damp for five to seven days after pouring helps the curing process. Do not allow it to dry out. Cover it with plastic, and dampen the surface twice a day or so.

In mid-July, clients are sometimes surprised to see me wearing long pants and long-sleeved shirts and gloves doing this work. I admit to sweating up a storm, but concrete is quite abrasive, and the Portland cement is highly alkaline, as caustic to skin as an acid. And Portland cement is hygroscopic, meaning it draws moisture from whatever it contacts — including your skin.

I hope this brief overview helps you understand this terrific tool a little better. **Figures 4-11** through **4-14** show the construction of a large pad and pier base for a 70 foot self supporting tower. **Figures 4-15** through **4-18** show construction of a base for a large crank-up tower using different materials. **Figures 4-19** and **4-20** show the much more modest base requirements for a Rohn 25G guyed tower.

Grounds and Grounding

Protecting your investment (tower, beams, feed lines, rotators and control lines, switchboxes, and the like) from lightning induced damage is a critical aspect of any installation. The basic idea, always, is to direct the current induced by the strike to ground as quickly as possible. Quickly means the path of least resistance — so the charge is dissipated before it enters the shack.

The idea is to provide a single, very low inductance ground (a system, if you will), so no current can flow "across" or through our gear, searching for that lowest potential. To do that, strapping works better than wire and

Figure 4-15 — KX8D needed to dig a large tower base hole as he had very sandy soil. A big hole goes a lot faster with big equipment to dig it. (KX8D photo)

Figure 4-16 — Because of the poor soil characteristics, KX8D needed to build a suitable footer to hold his concrete. (KX8D photo)

Figure 4-17 — The rebar cage for KX8D's foundation. (KX8D photo)

Figure 4-18 — KX8D's tower base after the concrete has been poured and the hole backfilled. This base hardware is for a crank-up tower. It attaches to bolts anchored in the concrete and is adjustable to make the tower level. (KX8D photo)

Figure 4-19 — This typical Rohn 25G base uses a buried base section. It's a lot smaller than the bases previously shown, which were for freestanding towers,

Figure 4-20 — WN3R's 60 foot Rohn 25G tower base. The top of the concrete is well above ground level, and cables to the station run through flexible plastic pipe which can be buried.

solid conductors are preferred over stranded. But even with the best protection system, some current can flow on feed lines, ac and control lines, so protection devices should be installed on these lines.

Starting at the tower base, it's wise to install a series of ground rods attached to each tower leg with bonding wires. Separate these rods by twice their length. Eight-foot rods would be 16 feet apart, for example. Keeping these wires and rods a few inches under the surface of the soil is fine; in fact, it's a good idea. But clamped connections should be inspected periodically, a good argument against burying them.

Figure 4-21 — An example of the molds used for Cadweld products. Here it's used to secure part of AB7E's tower ground system. (AB7E photo)

Figure 4-22 — A typical installation: The ground rods for WA4DT's new crank-up installation uses #6 AWG solid copper wire welded to a 10 foot, ⅝ inch copper-clad steel rod.

I like to use Cadweld One Shots (**Figures 4-21** and **4-22**) as a convenient way to attach the bonding wires to the ground rods, although strapping can be used with appropriate clamps. Here's a tip gained from experience: Igniting the One Shots can be troublesome. An ordinary torch won't work, as unused propane gas forms a "cone" around the flame, which prevents the ignition material from starting. So, I simply use an ordinary children's fireworks sparkler to light the One Shot. Cadwelded connections won't require inspection as often as clamped connections.

The next step is where many hams get into trouble — bonding those wires to their tower. When bonding copper to galvanized steel (or aluminum) tower legs, you *must* insert something between these two metals to prevent a galvanic reaction. I like to use stainless steel shim stock, for instance.

Care must also be taken in cleaning each surface. It's important to use a fresh cleaning pad or material on each surface to avoid contaminating them. It's also important to use a light coating of Penetrox or similar conductive paste on the joint before joining them. Use the properly sized clamps on the tower legs and tighten them very well. If possible, allow them to go through a temperature cycle (overnight, for example, then come back and re-tighten again.

Protection for Cables

When the tower takes a lightning strike (notice we're not saying "if" here), much of the energy can be induced into the feed lines and control lines, which will typically be taped or secured to the tower legs. To reduce the potential, the feed line(s) should be bonded to the tower. Typically, they're grounded at both the top and bottom. Towers taller than 100 feet should have multiple bonds. Most of the feed lines used by hams will have either bare copper or tinned copper jackets for the shield.

The standard grounding kits available from PolyPhaser, Andrew and other manufacturers, will have the necessary shims and conductive paste to ensure solid, reliable connections. The PolyPhaser kit contains lots of illustrations to guide you through the installation (use part number PPC-UNI-KIT-2CT, for copper shield to galvanized tower). It's important to note here that all of these connections are mechanical in nature. You do not want to use soldered connections anywhere. While fine for an RF ground, soldered connections will simply evaporate from the intense heat in any lightning strike.

The ground strap bonds to the tower itself will use stainless steel hose clamps, with both stainless band and screw. Or if you have an angle-leg tower, the ground strap will likely use

existing hardware holes. Again, clean all surfaces and use conductive paste.

PolyPhaser, ICE and Array Solutions all offer protection devices for rotator control cables. These are shunt-type protection devices, going to ground if a surge exceeds 82 V dc. Mounted to a tower leg just above the tower point ground, these are also useful devices to have in your protection arsenal.

Moving to the shack entrance, it's wise to install a bulkhead type panel (aluminum, copper, or stainless are all typically used), which simplifies the mounting of a variety of lightning arrestors and allows a simple method for attaching the copper strap to a good ground.

The arrestors should be mounted in a weatherproof box. PolyPhaser IS-50-series protectors are 50 Ω devices designed to mount on that bulkhead. Note that they are directional — antenna and equipment ports are clearly marked, and installing them backward will not provide protection. Models with UHF and N-style connectors are available, as well as a 3 kW unit.

For installations where the tower is located some distance (over 100 feet) from the shack, I prefer to have the IS-50 devices located both at the tower base, and then again at the bulkhead panel.

This is a good place to talk about the actual shack location itself. Basement locations are best — from a lightning protection point of view, not necessarily just for aesthetics, economics, or family relations. But the basement's closeness to ground can enhance protection possibilities. First floor locations are the next obvious choice, and shack locations going ever higher provide lessened protection potentials.

This overview introduces you to the basics of grounding — a systems approach that's truly necessary if your installation is to survive the rigors of a lightning strike. It's a wise approach to follow, providing not only some "insurance," but protection for your home, as well. For further details, there is a lot of useful information on the Poly-Phaser Web site.

A Few Words About Guy Anchors

Holes are also needed for guy anchors, as well, especially if we follow the guidelines provided in the Rohn catalog for a version of the "dead man" anchor. As shown in **Figure 4-23**, that's basically a large chunk of concrete suitably buried in the ground. The Rohn catalog contains drawings and specifications for anchors suitable for typical 25G, 45G and 55G towers. You can't go wrong by installing a concrete anchor following the manufacturer's guidelines. The techniques for building tower bases — digging holes, constructing rebar cages, pouring concrete — are the same for dead man guy anchors.

Questions invariably arise on ham reflectors and in correspondence, concerning other types of anchors — the ubiquitous screw-in earth anchor, elevated guy post anchors, anchoring to trees, and sometimes, even "temporary" anchors, which I have to assume means the tower installation will be temporary, say for a Field Day outing or something similar. In any case, let's examine some of these.

Screw-In Anchors

Screw in earth anchors (**Figure 4-24**) are widely used in the commercial world, anchoring utility poles, for example. There are a vast variety of them out there, ranging from simple, short shafted, small screw augers, to serious, heavy-duty, large screw anchors. The smaller anchors are *not* suited to hold up your tower, to put it bluntly. But, once upon a time, even the Rohn catalog carried a simple drawing of two guys, screwing in one of Rohn's own earth anchors, which were rather small — only four feet in length, with a six-inch helical screw.

The larger, heavy-duty anchors, especially those made by AB Chance, designed to be installed with a hydraulic driver, will hold up a tower — provided you have something akin to "normal" soil. Chance offers a series of anchors, along with a wealth of data on the holding power of such screw-in augers, on their Web site. Look for the AB Chance *Encyclopedia of Anchoring*, available for download in PDF format.

Rating these anchors is highly dependent on the properties of the soil. **Table 4-2** shows a sampling of AB Chance's ratings for some screw-anchor models used in Class 6 soil (defined as "loose to medium dense fine to coarse sands to stiff clays and silts loose sandy soil

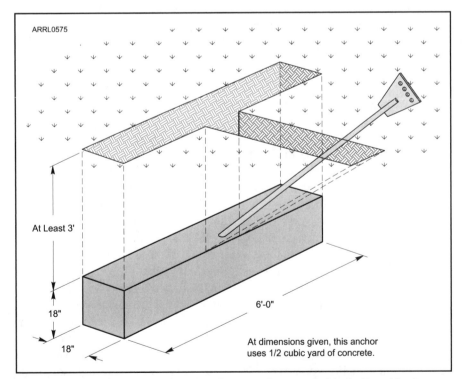

Figure 4-23 — A "dead man" anchor for guyed towers is basically a block of concrete buried in the ground. Block dimensions are specified by the manufacturer or calculated by an engineer and will vary depending on the height of the tower and wind and ice loading requirements. Various anchor rods are available, depending on the length and strength needed.

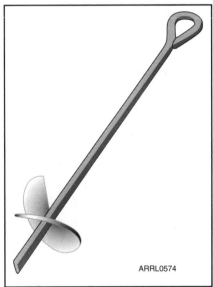

Figure 4-24 — Typical screw-in anchor construction. Pullout strength is determined by soil characteristics, anchor rod length and helical screw diameter.

or fill"). Dense sand or gravel or soils with higher clay content will have higher strength; loose sandy soil or fill will have lower strength. See the *Encyclopedia of Anchoring* for details.

There's obviously a direct correlation between the screw diameter, depth and pullout strength. What's a suitable screw-in anchor for ham radio tower use? Something with at least a ⅝-inch shaft, welded helical screws, and at least 5 to 6 feet long, are some of the things you should look for when searching various suppliers or catalogs — but check the ratings, consult with an engineer and consider having your soil tested. Longer, larger, bigger are all, of course, good attributes, although these very features will make getting the anchors in the ground more difficult, as well.

The heaviest duty models have two or even three helical screws on a 7 foot shaft, are professionally installed by machine, and are capable of handling huge loads. The AB Chance square shaft anchors (professionally installed) were used extensively at K4JA's installation of his 160 meter 4-square vertical array in Callao, Virginia. Pictures and brief captions can still be found at **www.k4za.com**. I was very impressed by them when I dismantled Paul's station.

I have screwed 4-inch screw helical anchors in to the NC clay soil by hand, but I have also bent a solid ¾-inch steel bar *attempting* to do that, too. Moistening the soil helps, but it takes an amazing amount of water to do that, along with considerable time for the water to soak down to where the auger is located. Although you'll read about this on the Internet or hear about it, the procedure should not be practiced.

The strength or holding power of such screw in anchors is derived from the mass of the earth resisting being pulled out of the ground. I've had clients tell me they drilled a hole and

Table 4-2
AB Chance Screw-In Anchor Ratings

Model	Screw Diameter (inches)	Shaft Length (inches)	Pullout (pounds)*
4345	4	54	3000
6346	6	66	5000
PS816	8	66	9000
10146	10	66	10,000
12537	14	96	15,000
012690AE	8 & 10	84	23,000

*For Class 6 soil. Table data presented for discussion only. Consult the AB Chance catalog and a professional engineer for specific design information.

Figure 4-25 — This interesting device is an expanding earth anchor. It fits over the anchor rod, is placed in a properly sized and angled hole, and pounded until it expands into undisturbed earth.

filled it with concrete, using the same screw in anchor they were unsuccessful at screwing in the ground. That's not an acceptable solution either, as that small cylinder of concrete can easily pull out of the ground with only 1000 pounds of pull. I know — I've tried it! In any case, it's a good idea to have the installer test a screw-in anchor to make sure that it works with the expected load. That will avoid surprises later on.

If you cannot get the anchor screwed to specifications, it's best to simply revert to the dead man style of concrete anchor. It's also true that you should be leery of any anchor you can *easily* screw in, as well. If it goes in easily, it may come out equally easily! Soils that contain lots of moisture (near creek beds or ponds, for instance)

should be avoided. Wet soil can slip around the screw. Any screw anchor's shaft should be marked where it leaves the ground so you can tell at a glance whether it's moved — especially during the first month or so after installation.

Expanding Earth Anchor

Another type of anchor worth your attention is what's known as the "expanding earth anchor" shown in **Figure 4-25** (AB Chance calls it a "Bust" Expanding Anchor). These bell-shaped anchors require you to drill a deep hole, at an angle or in series with the guy line. Then, the anchor is pushed down to the bottom of the hole by the anchor rod. A suitable length of pipe is then slid down, over that rod, and then smashed with a sledge hammer! You will feel (and hear) the anchor "bust open" as the expanding fingers extend out into undisturbed soil. Remove the pipe, then backfill the hole, tamping the dirt down as you work your way upward. Then, simply begin attaching the guy wires, as needed.

These anchors are terrific to use in problem areas and provide superb holding power. Anchor rods are available in lengths from 5 to 10 feet and diameters from ½ to 1¼ inches. For example, in Class 6 soil a 6-inch expanding anchor with ⅝-inch rod is rated to hold 8500 pounds. An 8-inch anchor with ⅝ or ¾-inch rod is rated for 15,000 pounds.

Many of these screw in anchors and the shaft rods for the expanding earth anchors have a large "knuckle" head on the end. The diameter of this head is large enough to accept guy cable without an extra thimble being added, but it typically only accepts two such lines. If you wish to install more than that, you can add an equalizer plate with a suitably sized shackle. Guy wire hardware is covered in detail in **Chapter 5**.

The Tower Itself

Planning the Work

With the base and guy anchors installed, it's time to take a close look at the tower itself and the tools required to work with it and on it.

Planning the Work

Each day's tower work always starts the same way — with the "tailgate meeting," which means the crew (climbers and ground crew) gathers around the truck's tailgate and discusses the work that's about to happen. This meeting is vitally important. It ensures that everyone has a good idea of what's going on, what's needed, what steps everyone will follow, what will happen first, and so on.

Do not attempt to do anything without having such a simple meeting. If you're a ground crew member and you have a question, don't be afraid to ask. If you're a climber, don't be afraid to ask or tell the ground crew what you expect or will need. The idea is to communicate the sequence of events for the day's work clearly enough so there are no surprises, let alone accidents. Not only will the climbers sort out their needed tools (which ones they'll climb with, which ones they will have the ground crew haul up, and so on), they'll plan out the jobs that need to be performed.

Someone will be designated as a safety person, the "just in case" guy with the working cell phone. Perhaps someone will be taking photographs. Make sure that doesn't interfere with the work. How the climbers and ground crew communicate with each other will be discussed — whether radios will be used, or other alternatives, and so on. Again, the issue is to have no surprises, and no confusion, disrupting the workflow.

Figure 5-1 — This Rohn 55 tower is slowly rusting away in the salt air in Aruba. Although this is an extreme case, it highlights the need for careful examination of any tower before climbing or dismantling it. (KK9A photo)

Preliminary Inspection

When we begin to seriously consider the actual tower we're going to put up, things get truly interesting. There are, as we have previously covered, two types of towers to consider — the guyed tower and the self-supporting tower.

Let's consider the guyed tower first. And let's consider the various kinds of construction used in them. We'll begin with tubular leg construction, as used in the popular Rohn G-series towers.

These triangular towers are assembled from steel tubing, with the legs or siderails held together through a series of Z-bracing rods, welded to that tubing at each intersection. Tower sections are typically 10 feet in length, and they bolt together with a five-inch overlap. The 25G series loses these five inches at each joint. The 45 and 55G-series are truly 10-foot sections, even after assembly. The G-series tower is the closest thing to a "universal" tower that we have in the USA.

The Rohn guyed towers (even the small 25G) are an amazing design — very strong for their size, weight and construction.

Hams are known for their ability to overload this tower beyond manufacturer's specifications and get away with it. A quick Google search will bring up any number of installation stories, pictures and guidelines.

There *are* some limitations, some possibly negative aspects to this design, which many hams tend to overlook. The most common issue is the ability to rust from the inside out, which is almost impossible to see or detect but is most certainly dangerous. Experience is probably the best defense here — knowing what you're looking at, knowing what you're dealing with and knowing when to walk away. In harsh environments such as salt air, rust damage can be obvious and severe (**Figure 5-1**).

It's quite common for the factory-applied hot dip galvanizing finish to last 20 years or more. It's quite common for there to be patches of rust at some point on the tower where something has been attached — a side mount, a rotator or a small antenna. These are usually stains from the associated hardware, and not a failure of the original finish. Tower sections that have been stored on the ground may suffer some deterioration to their finish from soil minerals and the extended exposure to ground moisture. Again, experience will be your best guide.

If the steel is still sound, having sections re-galvanized is an option in locations where you have access to such a facility. The sections are dipped in an acid bath for cleaning, and then re-galvanizing can take place. Typically, you pay "by the pound" to have such work done. It can be a cost effective way to restore sections and guarantee the life of your tower, whether it is tubular or angle iron construction.

For smaller jobs there are "cold galvanizing" coatings you can apply. These are typically spray-on applications. The better brands (ZRC, BriteKote) are zinc rich and require repeated shaking to keep the nozzle clear. Multiple, thin coats work best. It is *not* necessary to use a grinder, wire brush, or other abrasive tool on the section before application. Simply ensure the steel is clean, grease-free, and dry.

Takedowns

Hams are known for thrift — you see this at any and all hamfests, hear about it on the air and read about in the pages of any ham magazine. Nothing new there. But sometimes, it's so pervasive, I think we all must possess some specific genetic makeup that allows us to pass the license exam.

Normally, this isn't an issue. Indeed, it's somewhat a defining quality — part of the indefinable magic that makes this hobby so much fun. But when it comes to issues that can threaten life or property, economy has to be forgotten. Safety, security and seriousness must prevail.

A used tower can be a bargain. But it can also be an accident waiting to happen. Again, if you're considering taking down a tower for later use at your own station, seek out not only some advice, but also some real help from someone who's done this sort of thing before. Tower work isn't for beginners. And while it's impossible to gain experience simply by reading about it, climbing a tower to do a takedown is dangerous territory for a beginner. We all have to start somewhere, sometime, somehow — but it's best not to begin by taking down an existing tower with an unknown set of issues.

Temporary Guys

Let's assume you're taking down what appears to be a simple, easy job — some 25G tower. The most common mistake is *not to* install temporary guys or to use rope instead of proper guy wire as temporary guys. I learned this lesson the hard way. I once had 30 feet of 25G fall over with me attached to it, along with some temporary rope guys. Within seconds, I found myself laying flat on my back, still clutching the siderails, still belted in, but staring up at the bright blue Maryland sky, unable to breathe. For a couple of minutes, my mind was racing:

Wow, what happened? Why can't I breathe? Am I dying? How did the rope break? I can move my fingers and toes, why can't I breathe?

Actually, this printed description doesn't begin to come close to what I was truly thinking and feeling, but this is, after all, a family publication, for general audiences. Suffice it to say, I was scared witless.

Fortunately, the tower fell when I'd only reached the 25 foot mark. The landing spot was on a perfectly flat, manicured lawn, and the tower was just 25G, light enough not to seriously injure or kill me. The ropes did not fail. Two of them merely stretched out completely — enough so that the third rope (the side of the tower I was climbing on) simply collapsed or folded up. The pier pin base of the tower was loose enough to allow the plate to shift enough so the tower could lean, then fall.

In a few short minutes, I regained my breath, and we were ready to stand the tower up again. This time, we installed some real, steel guys, and used some more help to raise it upright. I will never again work on another tower with temporary guys made from rope. Yet, you read and hear about this practice all the time, especially regarding taking down the typical Rohn guyed tower.

Yes, it's somewhat troublesome to install, hook up, and secure temporary steel guys. But the simple fact of the matter is, I believe my life (and that of my co-workers on the ground) to be worth that trouble!

Time Marches On

The second aspect of taking things down that you must consider is the march of time — things will have changed considerably at the tower site since it was first erected. Some changes may be minor; some of them may be very serious. You must consider such changes, and deal with them accordingly.

Rust and the loss of steel should be your primary concern, as these will affect your safety. Then, ask yourself: Do you have space or room to allow the takedown to proceed without en-

Figure 5-2 — Rohn 45 base legs, uninspected for 30 years, rusted through on two siderails.

dangering yourself and the crew? Do you have room to lower or tram down antennas and other things mounted on the tower? (Nature has a way of closing in on the tower site and taking the land back!) ! Some of the North Carolina locals enjoyed a shot of me with a section of 45G hung up in an oak at N4XO's home, where the growth did not allow quite enough room to lower dismantled tower sections easily.

Loss of steel especially applies to the base of the tower and the guy wire anchors. You *must* inspect them, and carefully! It's wise to have a digging tool (I carry a military surplus entrenching tool with me), to allow you to remove some soil at and around the tower base, and also at and around each of the guy anchors. Make sure the steel has not eroded so much that it's now dangerous. I've encountered Rohn 45G with the siderails and the Z-bracing totally rusted through after 30 years of neglect (**Figure 5-2**).

Guyed Towers

Once you've determined there are no issues or problems, and the tower is safe to climb, you can do so. Cut cable ties and/or tape on the control lines and feed lines as you go up. Inspect each section joint, and the tower in general, as you make your climb.

Once at the top, it's time to decide how you'll proceed with taking things down. Multiple (stacked) antennas on the mast will require you to dismantle them one at a time, working your way up. That means you must take down the lowest beam, then perhaps remove the rotator (after securing the mast), and then lower the mast to allow access to

The TowerJack Tool

Working with Rohn's popular G-series tower (20G, 25G, 45G, 55G), is relatively easy. Most of the time. With some attention to details (such as pre-assembling sections on the ground, checking legs to ensure they're clear and that the bolt holes are free of slag from galvanizing), most tower assembly proceeds fairly quickly.

After the tower has been up a few years or even for a longer time, it will usually come apart easily and can be taken down without many problems. But sometimes…there's that one, stubborn, sticking section…that simply refuses to come apart.

Many of us have carried up small scissors jacks or hydraulic jacks, with blocks of wood and small ropes to secure everything, and struggled with taking sections apart. Sometimes, we also had problems getting them to mate, when putting the tower up. We accepted this, as simply standard procedure with this type tower, and went on.

Enter the TowerJack, invented by Jeanene Gill, N5UHL, and marketed by Jeanene and her husband Kenny, WB5HLZ. The TowerJack is a basic lever-type device (made of hinged bar stock), with hooks and notches located at appropriate points to facilitate assembly or disassembly of Rohn 20G through 55G sections (depending on which model you need). See **Figure 5-A**.

This simple lever will save you from working so hard when putting up or taking down such sections. The force this lever exerts is significant — yet another appreciation of some simple laws of physics. The only caution needed is to keep the jack to the left or right edge of the Z-bracing to prevent bending the steel bracing.

Trying to describe the process of using the TowerJack makes it seem overly complicated. In use, you either use a hook or notch, and you either pull down or push up on the lever handle. It's just that simple. It's a very clever, and very practical tool to have for anyone working with this style of Rohn tower. Indispensable, really.

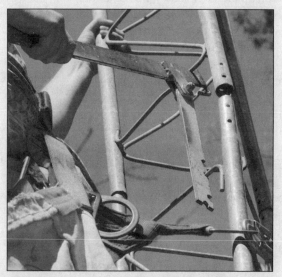

Figure 5-A — The TowerJack tool makes separation of Rohn G series sections easier.

and removal of the next beam, and so on. These are *not* jobs for the faint at heart, or for inexperienced or first-time climbers, in my opinion.

When the cables, antennas, mast and rotator are on the ground, it's time to consider taking the tower apart. This is another place where experience counts! You'll have to remove the guys to take the sections apart. Depending upon how and where the tower is guyed you may find yourself working on 30 feet of unguyed tower above the next lower set of guys. Temporary guys are

always a good idea. And those temporary guys should be steel — just like the permanent guys. No rope!

Yes, installing temporary guys slows things down, but I'd rather drive myself home at the end of the day and not be taken away from the jobsite in an ambulance. One of the nice things about ham radio is that it's typically *not* driven by a time clock.

Take your time, and work safely.

The proper tools are more important in dismantling and taking antennas and towers down than putting them up, I believe. Acquire a TowerJack tool (see the sidebar) for separating tower sections, Klein grips for grabbing onto steel EHS guy wire and some PB Blaster to soak and loosen rusty hardware, along with some suitably sized wrenches, and you're good to go in taking down most anything in the guyed ham radio tower category.

Self-Supporting and Crank-Up Towers

Self-supporting towers need to be inspected and cleared of cables, antennas and perhaps mast and rotator just like guyed towers. The major change in working with larger self-supporting towers is simply scale. Big, self-supporting towers will probably be taken down using a crane — the entire tower in one pick, after removing the antennas, for example. Small self-supporting towers, like the BX-series, require closer attention. Their steel is smaller, and their riveted joints can come loose over time. Usually, if at all possible, I also prefer taking them down with a manlift, in sections, or in one pick, after having removed the antennas. Working with cranes and manlifts is discussed in detail in Chapter 7.

Taking down crank up towers can be even more difficult. You now have moving parts to consider, parts that may have suffered from years of neglect or inattention and that may be in dangerous condition. The cardinal rule: *Never Climb A Crank Up Tower*, certainly applies even more when it comes to used crank ups. Again, I prefer to use a manlift. Take off the antenna(s), and then lower the tower to the ground. For those models with an erection fixture, pay particularly close attention to the winch and cable used before attempting to use it to lower the tower to the ground. I was quite surprised, once upon a time, to find that the center portion of the hinge pin on an EZ Way tower had simply dissolved into rust after 30 years!

Safety is always the watchword, and this fact is never truer than when working with used tower. Work slowly, paying particular attention to those parts that can wear or weaken, and plan your work.

Tools

There are some tools that are unique to tower work, and I'll spend some time talking about them here. Before that, let me simply say that the usual shade tree mechanic or handyman's selection of wrenches and assorted shop tools will cover probably 95% of the work associated with towers. There are any number of brand names — Craftsman tools are fine, for example. **Figure 5-3** shows part of what I'd consider a basic tower work toolkit:

- Deep well socket wrench set — ⅜-inch or ½-inch drive, sockets from ⁷⁄₁₆-inch on up
- Combination wrenches — ⁷⁄₁₆-inch to ¹¹⁄₁₆-inch (open end and box end)
- Screwdrivers — both #2 Phillips and straight blade
- Nut driver set — quality ones, with rubber grips
- Pliers — Channel-Lock, Robo-Grip or Klein lineman style
- Needle-nose pliers — larger size preferred
- Diagonal cutters

- Drift pin punch — taper punch (¼-inch on up)
- Adjustable wrench — 10 inch
- Knife — adjustable or retracting blade style preferred
- Hammer — something like a small-handled 9 pound sledge-hammer
- Klein grip — the only tool designed to hold EHS guy cable
- ComeAlong — used to take up or loosen that EHS guy cable

That simple selection should get you started. Over time, if you continue to work on your own installation, you'll grow accustomed to or prefer some tools to others, and you'll develop your own personal choices.

Now we'll get into some more specialized tools and accessories that I've found indispensable for tower work.

Climbing Accessories

Today's well-equipped climber takes some cues from the rock-climbing folks and utilizes some of the unique and specialized tools developed by and for them. **Table 5-1** lists several sources for climbing equipment discussed here.

Figure 5-3 — Tower work requires a good selection of hand tools. The minimum assortment includes ⁷⁄₁₆, ½ and ⁹⁄₁₆ inch wrenches (both box end and deep well sockets), wire strippers, pliers and screwdrivers. With these tools you can probably do 75% of on-tower tasks.

Why climbing gear, anyway? Primarily because it's lightweight and designed to safely hold or carry human loads, so the safety factor in helping ferry your antenna or rotator or tools to the top of the tower is significant.

Carabiners

The primary tool, of course, is the *carabiner*. Carabiners are lightweight metal links (usually aluminum) that can be closed and opened quickly and easily. Climbers, parasail gliders, spelunkers and rescue personnel use them in a wide variety of tasks. They come in a range of styles, sizes and designs. **Figure 5-4** shows a couple from my collection. For a good reference and starting point, look at this Web site: **www.uhartrescue.com**.

In typical climbing use, carabiners have both a connecting and a safety function. Connecting to a rope, connecting a rope to another piece of gear or connecting a rope to a fall protection device are primary uses. Carabiners must withstand extreme forces, as a climber's life often depends on them, so it's probably safe to clip your tool bag to the tower using one. As hams, we'll be more interested in the convenience factor — I never rely on a carbineer for any protective use.

But, as in climbing, different activities and tasks require different kinds of carabiners. Characteristics such as shape, gate type, strength, material, weight and size will determine the right type of carabiner for each task and budget. In choosing carabiners, simply look at the intended use first. If you can, buy a few carabiners of different styles to see which you prefer. If your budget is restricted, the simple designs (oval, non-locking) will probably cover most of your needs, although size can be an issue. The standard size carabiners will not fit larger-sized towers, for instance, like AB-105 or SSV towers. For that, you'll shift to the larger, rescue-work-oriented types (**Figure 5-5**).

Knowing how to clip a carabiner is extremely important. Basically, the bottom gate should always face away from the direction you are working. Second, your rope should never run across the carabiner gate. And when clipping, the rope must end up running out of the carabiner toward the climber. Finally, a carabiner must rest evenly against some support; uneven pressure on part of the carabiner will reduce its strength. (Again, in normal tower work, these factors are not an issue — most of our towers are not situated among rocks or boulders.)

Pulleys

Pulleys are used on every tower job. Mounted at the top of the tower as well as the base, they make hauling up tools, equipment or antennas relatively easy. You will find a variety of pulleys at any well-equipped hardware store, but these steel models are heavy. Tower climbers are always searching for ways to make things lighter.

I have two styles I use. One is a plastic model (**Figure 5-6**), rated at 1000 pounds, which has a hinged opening that allows easy access. Rescue pulleys are the next consideration from the

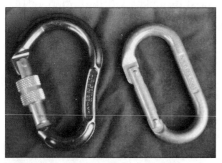

Figure 5-4 — Carabiners come in handy on the tower. I carry several locking (left) and simple non-locking oval carabiners whenever I climb.

Figure 5-5 — This larger carabiner works on ladders or larger towers such as AB-105.

Figure 5-6 — I like these rescue pulleys and associated carabiners for working with rope. The device at the upper left is a Petzl swivel (see text).

Figure 5-7 — The rescue pulley rotates, allowing it to be attached to the line at any point and not just the end.

rock climbing toolkit worth the tower climber's attention (**Figure 5-7**). These aluminum-bodied gadgets are ideal for tower work. They come in a variety of sizes, with ultra-smooth bearings, and best of all, they swivel apart, enabling them to be put on a line at any point along its length. Look for pulleys with a minimum 2-inch sheave size—that's the grooved wheel that spins between the pulley halves. Anything less will contribute too much friction or drag on your pull ropes.

Several people have asked about a couple gadgets they've seen me using during tower jobs, namely, a simple swivel attached to a rope or carabiner. But it's not just an ordinary swivel — this is a Petzl ball bearing-equipped swivel that makes tramming beams or orienting a rope's direction smooth and painless. A sealed ball bearing makes the action on this swivel silky smooth, even after years of use. Use it any place where twisting might be a problem. The top of the swivel can hold up to three carabiners. Petzl swivels are pricey, but are an energy and time-saving addition to my toolbox.

I never climb without at least one rescue pulley. I also always use one at the base of the tower (with a swivel), allowing ground crew personnel to run the rope through it and back away from the tower. The second pulley allows them to handle the rope horizontally, which many people find easier, and they don't have to worry about whether or not the rope is going to snag on the ground or is clear of obstructions at the tower base. Also, once safely away from the tower base there is less chance that the crew will be hit by falling objects if the climber accidentally drops tools or hardware. And away from the tower base it's also easier to look up and see what the climber is doing.

Slings

Finally, a few thoughts on slings. These are typically loops of nylon that are sewn together (**Figure 5-8**). The "norm" or standard is 25 mm wide (about 1 inch) sewn slings, manufactured from French-made *Faureroux* webbing. These general-purpose slings have impressive strength, typically a rating of 25 kilonewtons (kN) and an actual strength usually over 30 kN. (1 kN = 224.8 lbs-force; 4000 lbs-force = 17.8 kN.) When the webbing is made into a sling and tested, a strength of around 25-30 kN can be obtained. Generally, a particular length of sling is made a certain color, to assist in distinguishing different lengths. (I have slings in a variety of lengths. The 4-foot length, for example, is a versatile sling—a good length for attaching to an antenna boom to hoist beams.)

The real magic of the sling is that it's like that simple kid's toy, the Chinese finger puzzle — where the tighter or harder you pull, the tighter and harder the little paper tube grips your fingers. Relaxing your hands makes the paper loose, and you can easily remove your fingers. Slings work the same way. A few wraps and you can hoist a mast or other vertical member without the hassle of bolting on a muffler clamp or other device on the end. Some folks wonder about the safety and security of slings, or why you can't use rope to do the same thing. It is the makeup of that special nylon material that uniquely qualifies the webbing as the winner — rope slips, and the slings hold fast.

If you're familiar with cranes, you've seen different versions of these climbing tools in use. "Open slings" are a length of webbing with a small loop at each end. "Closed slings" are the standard circular slings with a stitched joint. These slings are usually heavier and wider material, and are designed for lifting dead weights. The configurations are usually different as well. The important point to remember about each type is that the surface around which the sling is secured is critical to safety. If any sling is worked around a rough surface, a lower strength than normal for the webbing will be obtained. And, a square edge or threaded bar, for example, will drastically reduce any webbing's strength.

I enjoy the speed and ease of use of these climbing accessories — that's why they're in my toolbox. Next, here are some ideas and guidelines for safe lifting with slings.

Safe Lifting

There are four primary factors to consider before lifting a load:

1) The size, weight, and center of gravity (CG) of the load

2) The number of legs and the angle the sling makes with the horizontal line

3) The rated capacity of the sling

4) The history of care and usage of the sling

The center of gravity of an object is that point where the entire weight may be considered as concentrated. To make level lifts, the lifting connection point (from a rope or a hook from a crane or wherever) must be directly above the CG. Some small variations are usually permissible, but if the lifting point is too far to one side of the center of gravity, the object will tilt over.

As the angle formed by the sling leg and the horizon decreases, the rated capacity of your sling also decreases. In other words, the smaller the angle between the sling leg and the horizontal, the greater the stress on that sling, and the smaller the load it will safely support. Hams will not normally encounter loads that even approach the limit of the typical 1-inch or 2-inch web slings. But, know what you're working with —

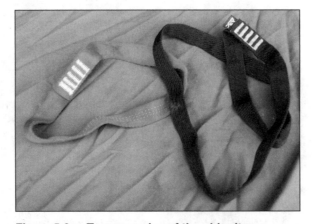

Figure 5-8 — Two examples of the ubiquitous climbing sling.

read the sling's label. The values listed are for new slings; older slings should be used with additional caution. *Never* exceed a sling's rated capacity!

The most common hitches we use to lift items are the *vertical hitch*, *choker hitch* and *basket hitch*. A vertical hitch is made directly from the lifting hook to the load, usually attached by means of another hook. A choker hitch means the sling passes entirely around the load with one loop passing through another on the sling's opposite end, forming a simple noose or "choke" hitch. The basket hitch means you pass the sling under and around the load, with both eyes or ends going back up to the lifting point. **Figure 5-9** shows a sling in use.

Sling angles are important. They directly impact the rated capacity of the sling. This angle is measured between horizontal and the sling leg, regardless of hitch. Whenever pull is exerted at an angle on a leg, the tension on the leg is increased. For example, each sling leg in a vertical basket hitch absorbs 500 pounds of stress from a 1000 pound load. That same load, lifted using a 60-degree basket hitch, exerts 577 pounds of tension on each sling leg. Without consulting some chart telling you what the increased load will be, follow this simple rule: If the *length* of

Figure 5-9 — Slings are a safe and convenient tool for rigging an antenna for hauling up a tower.

the sling leg is greater than the *span* or *width*, the lifting angle is okay.

You'll quickly realize that angles less than 45 degrees should be avoided. If you have trouble remembering, simply think about hoisting two buckets of water — one in each hand. It's very hard to lift and hold them with your arms outspread, but you can pick up and hold the buckets quite easily with your hands and arms by your sides. Keep your loads as vertical as possible, that's the idea. **Table 5-2** shows the effects of angle on sling capacity. Note that at some angles, rated sling capacity is cut in half!

Slings are swell tools, and they can make lifting our sometimes heavy and cumbersome three-dimensional Yagis so much easier. Combined with some modern climbing accessories,

rigging and lifting can actually be fun. It always takes far longer to set the rigging than to actually do the work of raising the antennas!

Understanding slings and those climbing tools will make tramming antennas relatively easy. Whether it's a small tribander or some monster Yagi, hauling antennas up and into place can be one of those moments of applied science that awes and amazes even the experienced rigger. For further reading and more detailed rigging information, take some time to peruse the Web site **www.fdlake.com**.

Rope

Rope is one of those things we simply take for granted. It's a simple fact that if you're going to be doing tower and antenna work, you're going to need and be using rope. Whether it's for tag lines, haul ropes, or tramlines, there are literally dozens of options and guidelines you should consider.

You'll find rope made from natural materials today (manila, sisal and cotton are the most common), along with lots of synthetics (nylon, Dacron, polypropylene, and so on). You'll find lots of sizes, prices, claims and comments, too, regarding each and every one of these materials. **Figure 5-10** shows some of the rope I use.

Table 5-2
Effect of Sling Angle on Capacity

Actual Sling Capacity = Rated Capacity × Sling Angle Factor

Sling Angle (degrees)	Sling Angle Factor
90	1.000
85	0.996
80	0.985
75	0.966
70	0.940
65	0.906
60	0.866
55	0.819
50	0.766
45	0.707
40	0.643
35	0.574
30	0.500

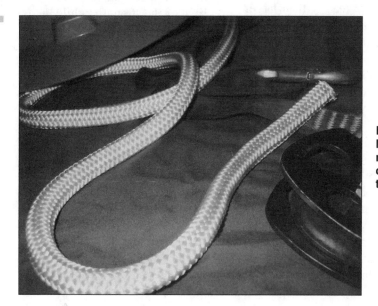

Figure 5-10 — Nylon braided rope is a good choice for tower work.

Size

Your first priority should be deciding exactly what task (or tasks) you want to *do* with your rope, as the use should dictate what you buy. Size is the first factor, of course, and this includes not only length, but also diameter. Material selection should be next. Price probably shouldn't be a factor, as once you start searching, you may be stunned by what you find. The old song, "you get what you pay for," is certainly true with rope, and thus, your budget for this item may require some rethinking. Good rope will be expensive.

Thinking about size is easy when speaking of length — you'll need to get up your tower and then back down, with a comfortable "handling" margin (say 25% to be sure) left over. So, if you have a 100-foot tower, you'll need a 225 to 250-foot rope.

Thinking about size is more difficult when speaking of diameter — you're into the arcane world of "working load" and "ultimate breaking strength," and almost no one understands these. If you plan on using your rope for lifting loads (tower sections, beams, and the like), then any rope having a working load of 100 to 300 pounds should handle almost anything a ham would commonly put up. Obviously, there are some nifty synthetic ropes out there of very small diameter strong enough to lift this load, but consider having (and using) something that will be comfortable in your grip. Smaller ropes hurt your hands more readily than larger ropes. Smaller lines (sometimes called cords) work great for "tag" lines, but once they're hauling a heavy load, they become difficult to hold. I'd say that a 3/8-inch line is the smallest size I would use for most any tower work. I prefer a larger, 1/2-inch rope for this work, myself.

Material and Construction

Thinking about material is also relatively easy, as the synthetic material's advantages outweigh their disadvantages. (All the natural fibers soak up water about as well as sponges do — they rot easily and must be stored properly at all times.) Nylon is the strongest

rope. It stretches, but because of this, nylon can absorb sudden shock loads that would cause other fibers to break. It's extremely resistant to wear, and can easily outlast natural fiber ropes. It's essentially rot proof, unaffected by most common chemicals, and most importantly, it knots easily.

All rope is constructed from small fibers either twisted or braided together. This twist, called the lay of the rope, is usually a simple three-twist construction, very common in any hardware store in America. Rope construction has evolved over the years, becoming more task-specific, and concerns with safety and good engineering practices from today's manufacturers provide several great choices. Braiding techniques, developed in WWII, should lead your rope selection process. Those choices are hollow braided, double braid and kernmantle ropes.

- Hollow braid means the fibers are laid around a hollow core in "maypole" fashion. As such, this rope is subject to flattening under strain, meaning it's no longer round.
- Double braid ropes are constructed with a "cover" over a "core," often of the same material. Rugged and smooth.
- Kernmantle ropes (*kern* means core; *mantle* means sheath) are usually made with an inner synthetic materials core, covered with a braided sheath. The core provides the strength, while the sheath protects that core from abrasion.

Double braided and kernmantle ropes will probably be the best overall value.

Strength

Now, what about working load, or limits, or strength? Obviously, we are now speaking of safety, so knowing the maximum safe working load for your rope can help keep you out of trouble. Never stress a rope (or line) anywhere near its breaking strength. As ropes age, or as they are spliced, stretched and subjected to sustained

loads or shock, or as they are exposed to ultraviolet light (in other words, as ropes are *used*), they will also lose some of their strength.

The rated breaking strength of seemingly-identical rope from different suppliers can vary by 10% or more, and different suppliers also specify a rope's safe working load at anything from 1/5 to as little as 1/15 of its breaking strength, so be sure to check on these specifications before buying. This is another good argument for using rope of 3/8-inch or larger diameter — as such rope has a high-enough breaking strength that even conservative calculations of a safe working load provide some leeway when working with typical amateur antennas.

This is a good point at which to talk about *working loads*, a term that is often misunderstood or misused. It's really impossible to posit an absolute safe working load for any rope or line. Calculations should be based on application, the conditions of use, and potential danger to personnel or property. You can determine a working load, along with some safety factor, using your judgment, personal experience and the manufacturer's published rating.

I always assess the risks involved when using a rope to do any task, taking that factor into consideration. Shock loading or heavy, sustained loads, and so forth, along with the physical factors previously mentioned, all contribute to a rope's service life. (The Cordage Institute considers the safe working limit to be derived by dividing the Minimum Tensile Strength of the rope by a safety factor, which ranges from 5 to 12 for noncritical uses. The factor rises to 15 when physical safety is an issue.)

Storage

Remembering that comment about exposure to ultraviolet light possibly weakening rope, it's important to store your ropes properly. You have to rely on your ropes, and damage from UV light or moisture may not be obvious. That's why I store all my ropes in Rubbermaid or other plastic tubs

Figure 5-11 — A plastic tub is a convenient way to store and transport the author's 600 feet of 9.5 mm climbing rope.

Gin Pole

The usual tool associated with pulleys and rope is the *gin pole*, a tower erection tool that consists of a long pole with a pulley at the top as shown in **Figure 5-12**. In every instance, the gin pole is designed to attach to the tower, with a haul rope passing through the pole, around the pulley (called a roosterhead, in the commercial world) and down to the ground. It's wise to install another pulley at the tower base, allowing ground crew workers to walk away from the tower, as they haul materials up the tower.

The principle of gin pole use is to secure the pole clamp to a section's tower leg, as high as physically possible, then slide the pole up as far as needed and secure the clamp. That usually locates the pulley at the top a foot or two higher than the work area, which is determined by the length of tower sections to be pulled into place and assembled.

The ground crew attaches the rope to the next tower section somewhere above the balance point and usually as high as possible without running out of work area at the top of the tower. The ground crew hauls the section up to the climber, who uses the "reach area" to maneuver the new tower section up and into place. **Figure 5-13** shows a gin pole in use.

When erecting tower sections, the next section is pulled past the top of the mating section — so the legs clear the top of the section — and then lowered on to those legs. This allows the most control, as the tower climber guides the mating sections together. This also ensures that the ground crew personnel always maintains "control" of the tower section attached to the gin pole, until it's fully mated.

A drift pin punch is used to align the holes, and then the bolts are pushed in and tightened. Then, the climber can go up, release the attached gin pole rope, collapse the pole, and move it up, on to this new section. The rope end is then sent back to earth, and the process is repeated, as necessary.

Rohn makes a gin pole specifically

(**Figure 5-11**). I simply lay them in, dry, when finished with a job. They always come out easily, too, which saves time on the next job. Each tub is labeled with information concerning that particular rope, usually prompting a question or two from my clients!

What Rope to Buy?

So, what should *you* do? What should *you* buy? Your decision will depend upon your ultimate use, the size of your tower, and how you intend to use that rope, of course. Obviously, having read this far, you'll realize you should probably have more than one rope, based on each of the previous points. For sheer strength, ease of use and working abilities, I'd recommend a double-braided nylon rope. I'd recommend a twisted (or better yet, braided) poly cord or rope for tag line uses. I recommend buying the best rope you can, then caring for it properly — sometimes more easily said than done.

Using rope as a tram to haul beams up the tower always generates considerable traffic on various Internet reflectors, mostly all related to safety, the laws of physics, and the right and wrong way to do such work. (Tramming is such a useful technique for raising antennas that it's covered in detail in Chapter 8.) My largest rope is a ⅝-inch double braided nylon. It's served me well, and the largest beam

I've ever hauled with it was a 200+ pound 20 meter Telrex monobander. If you're unsure of your needs or have questions about the requirements for strength or safety when working with rope, simply ask on one of the tower or antenna reflectors or bulletin boards. There are legions of people willing to help and supply the answers.

Buying rope can be expensive, especially good rope. With the increasing popularity of rock climbing, some interesting synthetic ropes can be found (be prepared, however, since all their dimensions are metric, as many ropes are made in Europe). Gear Express is one good source for climbing rope and hardware at considerable savings. (Mention your call, or what you intend to do with the rope; one of the owners is a ham.) Sterling Rope and Gear Shop and New River Nets are also good sources.

I wouldn't buy used rope. I wouldn't buy rope on eBay. I wouldn't borrow rope. I wouldn't suggest a club pool its resources to buy rope. I say this simply because the unknown variables increase exponentially, and it's not worth the risk. Why put yourself, anyone who's helping you, and/or the equipment you're working with, in any danger?

I consider my ropes an investment, and I have various sizes and diameters in my toolkit to allow me to work on towers up to 200 feet in height.

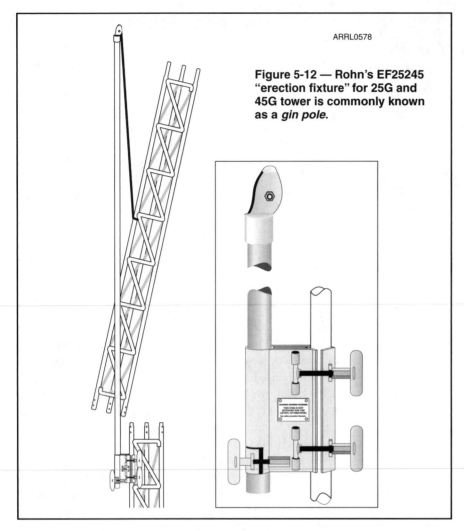

ARRL0578

Figure 5-12 — Rohn's EF25245 "erection fixture" for 25G and 45G tower is commonly known as a *gin pole*.

Figure 5-13 — A gin pole in operation. The ground crew has hauled the new section into position and the climber is guiding it into place onto the top of the existing one. (*N2VR photo*)

intended for use with their 25G and 45G models. Their designation is EF2545 (erection fixture). It is, however, very expensive. WBØW sells a gin pole with clamps suitable for both the G-series tower, as well as the BX-series. IIX Equipment also sell gin poles. The WBØW model is very rugged, and suitable for 55G, as well as the smaller towers. Lugging it around will tax even the biggest and best tower worker!

Folks often ask (in person, as well as on various Internet reflectors), exactly how to move the gin pole up, onto the next higher section. If I'm using the Rohn pole, I'll usually simply wrap a small sling around the pole, attach it to my belt and climb to the next attachment point. If I'm using a heavier duty version, I'll use a second work rope and allow the ground crew to pull the pole into position.

Gin Pole Wisdom (Rohn's EF2545)

The gin pole is your friend, just keep telling yourself that, as you struggle (especially the first time) to raise it or orient it on to a tower section right next to a guy bracket. It can be an awkward moment, and a seemingly cumbersome, clumsy tool.

The clamp is the important part, of course, and sliding the clamp to the end (where the pulley is) will allow you to maneuver the pole around the tower more easily. That's the first thing to remember.

I have climbed with the gin pole on my belt for years, but these days I often rig it to a sling and have the ground crew move it up and down, saving myself the struggle. On new installations, this requires climbing up and rigging the haul line beforehand, so it's sometimes easier just to move it myself.

Regardless, it's important to keep a clear path to the ground for the gin pole rope — so mounting the pole around or near those guy brackets can be problematic. The roosterhead (the pulley)

must clear the bracket as the pole is extended, for example, an often overlooked error when first using a gin pole.

The pulley itself should extend out and over, and in — toward the center of the tower — when lifting to install or remove sections. That position provides the best balance and the most lifting force. This means the haul line will come up one side of the tower, and down the opposite side. This provides plenty of clearance for the section, coming up or going down, too, to clear any guy brackets or guy wires that may already be in place.

Ease of use is obviously your first need on the tower, so it's always wise to store the clamp indoors, out of the weather, to prevent rusting. I always lubricate the clamp hinges before each use, too. The pole itself can be stored anywhere convenient, so long as it's free from getting bent, dented or damaged.

The gin pole can also be used to lift rotators or other tower accessories or other hardware up and into place. It can be cumbersome to use, though, and the pulleys described previously may be a better choice for hauling hardware. The gin pole works best for tower sections or masts where the pulley needs to be well above tower top and in line with the tower sections.

A Gin Pole Mod

While the Rohn gin pole is not exceedingly heavy, it can be awkward to use. The WBØW and W9IIX pole and clamps *are* heavy. So, in each case, a second climber, along with a short rope, can help you maneuver the pole up and into place on the next section. Or, a second work rope can be run down the opposite side of the tower sections, allowing the ground crew to lift and haul the gin pole up and into place as needed. A short sling, secured around the pole itself, makes attaching a rope and moving the pole much easier.

Having found myself atop towers numerous times, attempting to guide, insert, or otherwise install a long, heavy mast into a tower top or thrust bearing, I always found myself thinking, "There's got to be a better way to do this."

The problem is compounded by the gin pole's 12-foot length. A 24-foot mast is difficult to pick up and keep balanced — if you rig the "heavy end" down, the pole's too short. If you rig the "heavy end" up, no matter how much of a brute you are, the mast can get away from you. And yes, I know you can use ropes and turn the mast upside down, in the air. While there are, admittedly, other rigging solutions to the problem, I wanted something

Figure 5-B — The solution to long mast installation with a gin pole — the basketball hoop safety ring installed at the top. The hoop prevents the top-heavy mast from getting away from the climber.

a bit more elegant, along with a certain safety factor.

Driving home one afternoon, the solution came to mind as I watched some neighborhood kids playing basketball. Why not simply mount a hoop (like a basketball hoop, that is), at the top of the gin pole? Then, regardless of rigging, regardless of where the heavy mast's balance point falls, it cannot get away from the climber or ground crew as they work together to guide the mast into place.

A quick trip to my local sports emporium proved the plan to be economically feasible. I bought a hoop for $20. After purchasing it, and a few minutes with my trusty DeWalt grinder, I'd removed all the little welded or brazed-on hooks and loops designed to support the netting. I then drilled some holes to accept a couple of 2-inch U-bolts on the plate for mounting to the gin pole, and then sanded and repainted the whole hoop. **Figure 5-B** shows the result.

Naturally enough, on the next job requiring me to install 21 feet of Chrome-Moly mast inside a 100-foot Rohn 25G tower, my ground crew got quite a chuckle from this newest gadget in the K4ZA toolkit. Amidst cries of "Be like Mike," and "Go for three," and so forth, we slowly raised the heavy mast into place. But all the cries and catcalls ceased when the mast slid into place without incident, not only saving us time and energy, but also removing one of those lingering doubts about another dangerous job. It's a simple and elegant solution to this old problem.

The hoop is relatively light, inexpensive, and the climber can install and remove it in minutes. It's another welcome addition to my toolbox.

How-To Tips

Here are some tips for stacking tower sections that I've picked up over the years.

Checking each tower section carefully before attempting assembly makes perfect sense. This applies to new tower sections as well as used. Tubular legs must be clear of debris and obstructions. Look for excessive galvanizing and make sure that the bolts fit the holes. If a leg is blocked, sliding a ground rod through the siderail works well. It has enough "mass" to force out any debris, while still being small enough to easily slide through.

Pre-fitting the sections on the ground may seem overly cautious and/or troublesome, but it's far easier to find problems on the ground than while in the air. Granted, neither place may provide an easy, immediate solution, but the tower crew will appreciate and applaud your efforts. Even brand new sections may have been banged around in transit, bending the ends of the legs.

The TowerJack tool, ratchet straps, even long "breaker bars" have all been used to successfully shift stubborn legs into alignment. As a last resort solution, I've even bent the Z-bracing, which has the effect of moving the opposite-facing siderail leg. Beating it up or in toward the center of the section shifts the legs inward; beating it down or out shifts it outward. For those truly stubborn moments on the tower, I've also installed the mating section on the misaligned leg, and then used that section itself as a long lever arm to shift the leg where I needed it. These are, as I say, solutions that work, but shouldn't be tried by tower novices or those faint of heart.

Lubricating the legs is always a good idea. I like to use white lithium grease, mostly because it comes in a convenient-to-use tube. For those towers that will be used as 160 or 80 meter verticals, I'll use Penetrox or Noalox antioxidant compounds.

Tower bolts are controversial. I admit to using stainless steel in assembling Rohn tubular towers. Yes, this violates the "Do What The Manufacturer Says" rule. Yes, stainless is not as strong as other material. Yes, this makes disassembly much easier later on. And, I admit, I simply hate rust. I find it offensive to look up at a tower I've put up and see it stained with rust at each section joint, seemingly within a few short weeks after assembly. In 30 years of tower work, I've not encountered one problem with using stainless bolts.

Guys For Guys Who Have To Guy

I believe guy wires to be one of the most misunderstood pieces of tower hardware around. I say this because I've encountered everything from simple clothesline to PVC-jacketed aluminum CATV cable used to guy towers! (No, I did NOT climb the tower supported with the CATV cable. The owner was quite upset; assuring me that tower had gone through Hurricane Hugo without mishap.)

You need to consider the maximum load that will be put on the guys during operation, along with some safety margin. Consider maximum uplift forces the guy anchors must sustain. Having done this, you can then design or engineer a guy "system" based upon:

■ ANSI/EIA/TIA-222-G
■ Soil conditions
■ Wind load of everything that's going on the tower

It's important to think about what you "might do" one day and design a guy system that not only works today, but also later — if and when you add things to the tower. This *will* happen. In other words, you're not just putting up "50 feet of Rohn 25" or "100 feet of Rohn 45." You're also putting up a guying system.

Rohn and other manufacturers publish information about guy requirements for various tower configurations, loads and environmental conditions. **Figure 5-14** shows a couple of examples. If you need help in determining how to adequately support or guy your tower, the ARRL's Volunteer Consulting Engineer program can steer you to a knowledgeable engineer. **Table 5-3**

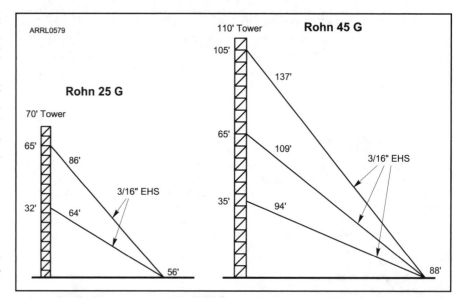

Figure 5-14 — The Rohn catalog includes guy specifications for many different tower configurations and wind speed ratings. Here are the dimensions for two typical ham towers — 70 feet of Rohn 25G and 110 feet of Rohn 45G. These designs are for TIA/EIA Standard Rev G at 90 mph or Rev F at 70 mph. Consult the Rohn catalog or a professional engineer for complete design information.

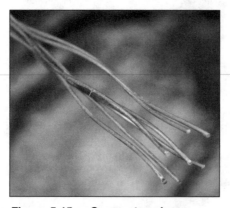

Figure 5-15 — Seven strands are visible in this piece of ³⁄₁₆ inch extra high strength (EHS) steel guy wire.

shows specifications for some of the guy cable materials discussed in the following sections.

EHS Steel Guy Wire

EHS (Extra High Strength) steel cable is every manufacturer's recommended guying cable. Sizes start at ³⁄₁₆ inch and go up, next to ¼-inch EHS, and then for truly tall ham towers in higher wind zones, ⁵⁄₁₆-inch or ³⁄₈-inch EHS. Most ham towers will use the ³⁄₁₆ or ¼ inch cable (**Figure 5-15**). EHS guy wire is stranded and each strand

Table 5-3
Guy Cable Comparisons

Cable	Nominal Diameter (inches)	Breaking Strength (lbs)	Weight (lbs/100 ft)
³⁄₁₆ inch 1×7 EHS	0.188	3990	7.3
¼ inch 1×7 EHS	0.25	6700	12.1
HPTG6700I	0.37	6700	5.0
⁵⁄₁₆ inch 1×7 EHS	0.313	11200	20.5
HPTG11200	0.44	11200	7.0
³⁄₈ inch Fiberglass Rod	0.375	13000	9.7

EHS steel cable information is taken from ASTM A 475-89, the industry standard specification for steel wire rope. The HPTG listings are for Phillystran aramid cables, and are based on the manufacturers' data sheets.

Figure 5-16 — The Klein grip (commonly called a Chicago grip) is the proper tool to hold EHS guy wire.

Figure 5-17 — Crosby Clips (left) versus hardware store cable clamps. The wide saddle provides much better gripping power, does not deform the cable, and makes the nuts more accessible for tightening.

is galvanized. It's tough stuff and requires some knowledge and special tools to work with.

The proper tools to use when working with EHS guy cable are Klein grips (**Figure 5-16**), specifically what's called a "Chicago" grip or the "Haven" grip. These tools allow you to grip the EHS securely, apply tension, and then attach the Preform or Crosby Clips, and associated hardware as described in the following sections.

The Chicago grip #1613-40 is the model to look for, as its parallel V-shaped jaws provide four long points of contact on the cable, allowing it to be tensioned until terminated.

The Haven grip is smaller and lighter, but its curved jaws grip the cable for only ¼-inch, possibly deforming it under tension. The #1604-20L is the model to look for, as it has a swinging latch to help hold the cable in the jaws.

CABLE CLAMPS

Attaching EHS cable to your tower and to the guy anchor is where the fun begins. The least expensive solution is to use what are commonly called *cable clamps*. Cable clamps have been around forever and you can find some at the local hardware store. Those hardware store clamps are usually made of malleable iron, and *not* forged material, and therefore are not that strong. They are prone to rusting quickly outdoors, and are thus not recommended for tower use. The first clue that hardware store

cable clamps are not suitable for tower guy wire applications should be the cheap price, usually less than a dollar!

The clamps we're talking about are specially designed for use with EHS guy wire. *Crosby Clips* is a popular brand name that you may have heard of, and that's the term I'll use here to distinguish the proper clamps from the hardware store variety. These clamps are heavy duty, made of drop forged construction and galvanized for long life out in the elements. **Figure 5-17** shows Crosby Clips and hardware store cable clamps side-by-side. The differences are subtle but important.

I no longer use wire rope clips or cable clamps except on some short ends of EHS guy cable used as safety wires on the turnbuckles, and so forth. I encounter them on the job, and they are often used incorrectly. Here's the proper method:

Refer to **Figure 5-18**. The clamp consists of a saddle and a U-shaped clip or saddle. The end of the guy cable is turned around a thimble and doubled back on itself, creating two

"ends." One of them is termed the *live end*, and the other termed the *dead end* (that's the short piece that terminates or ends past the thimble). Tightening cable clips to the manufacturer's required rating is done with a torque wrench.

Apply the first clip one base width from the dead end of the cable. Put the U-bolt over the dead end of the wire and the saddle over the live end ("never saddle a dead horse"). Tighten the nuts evenly, alternating from one nut to the other, until you reach the required torque. Apply the second clip as near the thimble loop as possible. Tighten the nuts appropriately, alternating again on the nuts, to the required torque. Put the third clip equal distance between the two, and tighten as before. Wire rope clips will loosen over time, so check them periodically.

The typical clip will deform the cable, so using a double saddle clamp makes sense. Crosby's brand name for them is "Fist Grip" clamps. They're especially useful in that they allow you to tighten them using a full arc with your wrench.

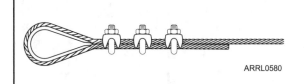

Figure 5-18 — Proper cable clamp termination. "Never saddle a dead horse" is a popular way to remember which side the saddle goes on.

ARRL0580

PREFORMED LINE PRODUCTS BIG GRIP DEAD-ENDS

The guy cable end fitting of choice these days in tower work is a device hams commonly call a *guy grip* or a *big grip* or a *preform*. Made by the Preformed Line Products Company (**www.preformed.com**), their official name is Guy Grip Dead-end or Big Grip Dead-end. These grips come in a variety of sizes (all color coded), designed to fit all the popular guy materials in use today. Grips for EHS guy wire sizes commonly used by hams are available from a number of vendors such as TESSCO or Texas Towers. **Figure 5-19** shows a Big Grip partially installed on EHS guy cable. (Big Grips are the Preforms we want to use in guying our towers. A standard or typical Guy Grip is just slightly shorter, and is intended for use by utility companies in guying their poles. The loads imposed by a wooden pole are different — much more "static" compared to the dynamic load imposed by a tower. That slightly longer length provides significantly more holding power!)

The Big Grip Dead-end works by applying an even and constant radial force on the strand. This force is achieved because the inside diameter of the grip is 80% of the outside diameter of the cable or strand. The gripping power also comes, in part, from the aluminum oxide coating on the grip's wraps (that's the crystalline coating on the wire).

The grip has a short side and a long side, and you begin wrapping using the short side. Begin with the end of the guy wire just coming past the paint marks. Wrap the short side completely. Then, insert or install the thimble, and begin wrapping the long side. You line up the first set of paint marks and continue wrapping, right to the end. The final wrap will be difficult, and a quick twist with a pair of pliers will probably be needed to "click" the wraps in to place. (I find that RoboGrip pliers work great for this use.) Another option is to split open the legs of the dead-end and wrap the resulting subsets individually, which requires less force to snap-in.

Figure 5-19 — This Preformed Line Products Big Grip Dead-end is partially installed on EHS guy wire.

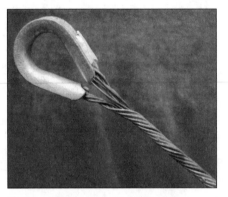

Figure 5-20 — Here the Big Grip is completely installed along with a proper heavy duty thimble.

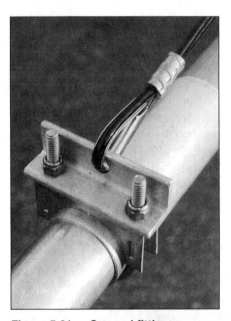

Figure 5-21 — Swaged fittings are often used on stainless steel cable or wire rope. They require a special tool to compress the fitting over the cable. Here, a Nicopress swaged fitting secures an element truss made of Phillystran cable on WN3R's MonstIR Yagi.

The second set of paint marks are used when the grip is passed through an insulator. Simply line them up, as you did when beginning to wrap the EHS. Once you use a PLP Guy Grip or Big Grip Dead-end, you will probably never want to go back to a cable clamp type grip. Ever! These grips are clearly the winning tool of choice. **Figure 5-20** shows a Big Grip installed with a thimble.

Questions often arise about the ability to "re-use" these grips, since they're so convenient and easy to use. And they can be applied three times, but only within a three-month time span. Any longer than that, and you must simply toss the grip out. And no, the grips should *not* be applied as "pulling tools" or used for other non-guying purposes.

Preformed Big Grip Dead-ends are also available to fit other guy material, such as various sizes of nonmetallic Phillystran and Polygon rod described later in this chapter.

SWAGED FITTINGS

Another end fitting is a swaged fitting, usually a Nicopress fitting. These low profile fittings look neat, and each crimp on the fitting has been determined to be as effective as a single cable clamp (typically, there are three crimps per fitting), so they're obviously also strong. They are relatively inexpensive, but the tool needed to install them is quite expensive. Once installed, they cannot be re-used, and therefore must be cut off. **Figure 5-21** shows a typical Nicopress fitting installed on a boom truss.

Guy Hardware

In addition to Crosby Clips or Big Grip Dead-ends, there are some other pieces of guy line hardware you should know about and use.

Thimbles and shackles should always be used with Big Grips — thimbles to assure the grip does not collapse, and shackles to allow you to move or remove or change the configuration without destroying the Big Grip. In short, the anchor shackles provide a method for easy maintenance. **Figure 5-22** shows some heavy duty shackles

Figure 5-22 — Anchor shackles are the hardware of choice for attaching guys.

suitable for tower use. Once again, a galvanized finish will provide many years of service.

A ³⁄₁₆-inch Big Grip requires a ⁷⁄₁₆ to ³⁄₈-inch thimble. A ¼-inch Big Grip requires a ½-inch thimble to properly match its bend radius. And, there are different kinds of thimbles. You do not want the "teardrop" shaped ones — with the closed end. You want heavy duty (HD) thimbles, suitably hot dip galvanized or otherwise protected from the elements. They should be U-shaped, expanded ones, with plenty of room to fit the Big Grip's end. Such thimbles will likely not be available at your local hardware store, but suppliers and vendors of tower hardware will have them. **Figure 5-23** shows a variety of thimbles useful in tower work. All this may seem overly critical and detailed, but any job worth doing is doing right.

At each guy anchor, you will need some way to adjust the tension on the guy cable. For that, we use a turnbuckle. But once again not just any ordinary hardware store item (**Figure 5-24**). (Indeed, hardly any hardware suitable for tower work will be found at a hardware store!) To support the load generated by a tower, you want something made from forged materials, not stamped or rolled, and galvanized or otherwise weatherproofed. You want something with at least nine inches of adjustment (the threaded part). You want either *eye-to-jaw* or *eye-to-eye* configurations. (See **Figure 5-25**.) The turnbuckles attach to the anchor and to the EHS guy with anchor shackles.

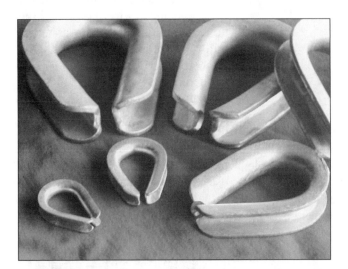

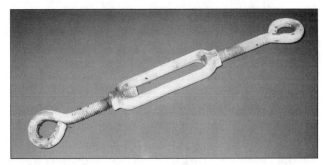

Figure 5-23 — Heavy duty thimbles are the only kind to use on guy wires. Note the rounded nose and deep grooves.

Figure 5-24 — Hardware store turnbuckles are not suitable for tower use. The eye can open or come apart under pressure.

Figure 5-25 — The "working ends" of ½ inch eye-to-eye and eye-to-jaw turnbuckles for tower use.

Figure 5-26 — An equalizer plate (EQ plate) is used to attach multiple guy wires to a single guy anchor.

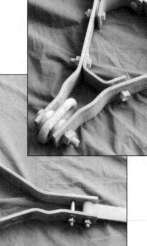

Figure 5-28 — The old style Rohn 25 guy bracket uses torque bars.

what's known as an equalizer plate, or EQ plate. (See **Figure 5-26**.) The EQ plate allows multiple guys to attach to the guy anchor.

As long as we're mentioning shackles and thimbles, it's time to mention guy brackets as well. I strongly recommend using them. While you can simply wrap the Big Grip around the tower leg (even the small Rohn 25's 1.25-inch diameter siderail is large enough to ensure the Big Grip doesn't collapse), this method of mounting does nothing to mitigate against tower torque induced by wind or the stopping and starting of the tower loads. Such twisting torque can be offset with the standard Rohn guy brackets, which transfer the forces to multiple tower legs. (Besides, why put something around the legs of the tower which will rub continuously and eventually remove the galvanizing?)

There are new style and old style brackets for Rohn G series towers (**Figures 5-27** and **5-28**). You'll no doubt encounter them both eventually. For the Rohn G-series, there's been lots of discussion regarding the use of torque arm guy brackets. Rohn actually discontinued this old style for a while, but "popular demand"

If you have only one guy line (on a small tower, for instance), you might choose to simply use the knuckle end of the GAR-30 Rohn ground anchor, attaching to it with an anchor shackle. But if you have a taller tower, and multiple guys, you might wonder or ask how you attach those multiple guy wires to the anchor. If you're really clever, you'll have realized that tightening the lower guy wire, for example, will then tighten the upper line. To help spread out or distribute these loads, the common way to attach them is to use

Determining Guy Wire Tension

Once you've got your tower up, and reasonably plumb, you'll inevitably ask yourself, as you turn away on your turnbuckles....*how tight do I make this?*

Proper guy wire tension is important for the life of the tower system. Stability is the key. You want to transfer wind forces to the guys and guy anchors. Too much tension, and you put excessive compressive forces on the tower legs.

Too little tension, and things move around, inevitably leading to other problems later on. After years of tower work I've learned that most folks with towers in their backyards have no idea what the tension actually is on their guy wires. I can also assure you most of these guy wires are too loose. So, what's the answer? Using a simple tool designed for marine use (to set tension on sailboat rigging), we can easily and accurately measure our guy wire tension. That tool is the LOOS tension gauge.

The LOOS gauge (model PT-2) measures 3/16 to 1/4 inch EHS guy wire easily. See **Figure 5-C**. Available at most marine shops or online, the gauge could easily be a club purchase or an addition to your own toolbox if you wish to include guy tension in your own yearly inspection plans. (Their model PT-3 measures up to 3/8-inch cable.)

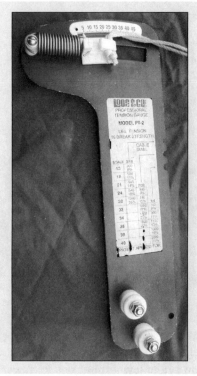

Figure 5-C — The LOOS gauge model PT-2 is used to set and adjust guy wire tension to specification.

Plumbing a Tower

One of the more misunderstand aspects of tower erection is making sure the tower is vertical, or plumb. As usual, there are a variety of methods available to you. Here are some thoughts on some of typical ways to ensure straightness.

The Transit Method

The choice of the professional, a transit is a special form of movable telescope mounted within two perpendicular axes — horizontal and vertical. Transits came into widespread use during 19th century surveying, allowing greater accuracy in laying out railroad lines in the American West for example.

A simple builder's transit (*not* a builder's level), which can measure both vertical and horizontal angles, will allow you to precisely line up your tower. They can often be rented locally. This is merely an abbreviated outline of how to use this tool to plumb your tower.

Using the transit takes some practice, but it's not difficult. See **Figure 5-D**. Pick a spot that's as far away as your tower is tall. Set up the tripod 100 feet back, if your tower is 100 feet high, for instance. Push the tripod's feet firmly into the soil, aiming to have the top plate more-or-less level. Then, attach the transit head to the tripod (keeping one hand on the instrument until it's secure!). Use the leveling screws, and level the instrument, checking through 360 degrees of rotation. You begin with the plate bubble parallel to the two leveling screws. Use the "left hand rule," meaning turning your left thumb to the left and your right thumb to the right, will move the bubble to the left. The inverse is also true, if you need to shift the bubble to the right. Once the transit is level, rotate the telescope 90 degrees. Check and re-level.

Rotate 90 degrees again. Continue this process until the plate bubble remains in the same position throughout the entire 360 degrees. Check the level bubble before each reading.

Now, it's a simple matter to line up the crosshairs on the tower leg near the bottom, lock the horizontal screws, loosen the vertical screws, and tilt the eyepiece up the tower leg. The first time you do this, you'll likely be amazed and think your tower is totally out of plumb. But you're looking at a magnified view in the eyepiece.

To totally check, move the transit 90 degrees away from your first location, and repeat the procedure. (You see why using *two* transits is faster?)

Adjust the guy wire turnbuckles accordingly to bring the tower into plumb.

Working Without a Transit

If you do not have access to a transit, it's possible to use a simple builder's or carpenter's level and achieve success. It's wise to use a longer level — a four-foot model, for example — in order to ensure accuracy. Plumb the base section before pouring the concrete. This requires holding the base section steady, of course. (I will often

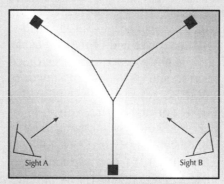

Figure 5-D — Where and how to orient a pair of transits to ensure a tower is straight and plumb.

mix and pour a bag or two of premixed concrete around the section to ensure the base doesn't move. This has always worked well for me.)

Once the base is plumb, the other bolted-together sections will follow the vertical path rather easily. Sighting up the assembled sections will immediately tell you if something is wrong.

A simple plumb bob hung from the tower will aid in alignment. And while I've heard about using piano wire and cement blocks, I've never found that necessary. A simple string plumb bob has always worked for me. If there's wind, and the bob will not "settle down," putting it inside a bucket of water placed at the tower base works. You can make your own plumb bob using string and a suitable weight. (I've used everything from a large nut to a hammer or wrench to a brick — the "line" is what provides your reference, not the "bob" or weight on the line's end.) Simply sight the line against one tower leg, then move around and sight it against another. This will be amazingly accurate.

Laser levels have come into widespread use within the past few years, and I own one. It is not as handy as I first thought it would be, but it worked great stuck on a tripod (once leveled) to establish the three anchor leg points for a large, tapered freestanding tower, replacing my old homemade water level. The problem with the laser is seeing it in bright sunlight much above 50 feet.

Clients often wonder (and ask) how much the tower can be out of plumb and still be okay. The EIA/TIA guideline for guyed towers, for example, allows no more than 1 part in 400, or three inches in 100 feet. For self supporting towers, the deviation allows no more than 1 part in 250, or about 4.8 inches for 199 feet. This level of accuracy is easy to obtain, if you begin with a level and plumb base!

---

seemed to bring them back. Personally, I've always liked using them. Here's why: the guy bracket design distributes the load evenly among the tower legs. The bracket provides a convenient and properly designed guy attachment point. And the bracket must provide *some* mechanical advantage. With Rohn 25, for example, the tower's radius will be just over seven

inches. That 12-inch torque bar will increase that radius by a bit more than 8.5 inches, surely a good thing. You can "feel" the difference.

Alternatives to Steel Guy Cable

We all know (and have seen at various venues) that the antenna pattern of

a Yagi can resemble a large donut surrounding our tower. Obviously, some of that RF must pass through the steel cable guy wires if we're using a guyed tower — exactly the case if we've put up a tall tower utilizing stacked Yagis, and so forth. Suddenly, the issue of interaction arises. We don't want interaction with guy wires to change the performance of our antennas, so the immediate question becomes: What to do? For years, hams have used insulators to break up steel guy wires into nonresonant lengths. The *ARRL Antenna Book* even includes a chart showing lengths to avoid. With the addition of 12, 17, 30 and 60 meters, finding a suitable length has gotten harder.

Nonconductive guy cable is the other alternative popular with hams. The choices are Phillystran (**Figure 5-29**), a flexible nonmetallic cable based on aramid fiber, or Polygon fiberglass rod. A quick perusal of the various available catalogs will show that both nonconductive choices are somewhat more costly when compared to steel. But is that the end of the story? Are there other factors to consider? Yes.

If you intend to break up EHS guy wires into nonresonant lengths with an appropriate number of insulators, you should consider the cost of labor, whether you hire someone or take your time away from other things. You should consider the cost of those insulators, which must be of a size and strength appropriate for the application. You should consider the cost of the extra Guy-Grip Dead-ends needed to attach the wire to the insulators

Figure 5-29 — Nonmetallic Phillystran guy cable is made of aramid fibers covered by a protective outer jacket.

(**Figure 5-30**). You should consider the time and effort required to haul up a heavy length of steel cable holding some heavy ceramic balls, undoubtedly more difficult than hauling up a simple hunk of lightweight Phillystran. To do justice to your installation, you need to sit down and do a line-by-line analysis of all the necessary materials and labor.

If you are building a large station, Polygon rod can be a cost-effective alternative, provided you can utilize the minimum order size of 5000 feet of the material. Group buys can be one way around this, but be aware the material (shipped rolled in 10-foot diameter reels), must be unrolled very soon after delivery in order to preserve its integrity. And the rod can be problematic to handle, requiring a simple mask, long sleeves and good gloves, to prevent the fibers from getting inside you!

It's important to remember that both Phillystran and Polygon rod are fragile — easily damaged by fire. They can be cut or scored with simple hand tools. For these reasons, these guy lines are typically not brought all the way to ground level. They receive steel "stingers" usually from 15 to 30 feet in length, which then connect to the guy anchor.

At the tower end of the guy cable, it's common practice (as well as a good idea) to add another short stinger of steel EHS guy wire at the guy bracket when working with Phillystran or Polygon Rod guy material. Simply install two Big Grips back-to-back. Then attach the nonconductive guy material and continue to the ground, or to the steel stinger on that end as described above.

Should something bad happen, such as a beam gets dropped during han-

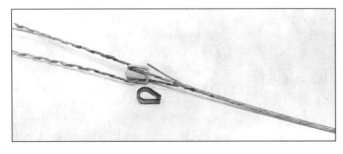

Figure 5-30 — Steel guy cables can be broken into nonresonant lengths by using insulators.

dling or fails and falls later on, damage would most likely occur close to the tower. Protecting nonconductive material from being crushed, cut or otherwise damaged from such an incident is a good insurance policy — both long term and while working on the tower too.

Preformed Line Products makes a series of Big Grip Dead-ends specifically for use with the various sizes of Phillystran and Polygon rod guy cables. Appropriate thimbles and anchor shackles should be used, as described previously.

House Brackets

House brackets are often used, especially with Rohn 25G or other small cross-section towers. The house bracket installation takes the place of a set of guy wires at some convenient height on a structure.

House brackets are often attached with lag screws, but a better solution is bolts that pass through the siding to thick "backer boards" mounted behind structural framing timbers inside the wall or ceiling. Remember, too, that lumber dries out over time. It often moves or changes size and sometimes loses strength or resiliency. That's a good argument for making the backing plate from steel instead of wood.

If you have any questions about how to secure a house bracket, consult your home builder, or a framing carpenter, or other structural expert before risking damage. The forces transmitted from a tower to your building can be considerable.

Remember, too, that the installation of the house bracket will "drive" the actual tower installation — the distance the house bracket's clamps extend out from the structure will determine tower plumb, and where and how the base should be situated.

Star Guyed Towers

Once again, Hank Lonberg, KR7X, provides us with a professional engineering perspective. This time the topic is star guys (using two guy wires per anchor point — **Figure 5-31**) ver-

Figure 5-31 — Star guys in use at NR5M.

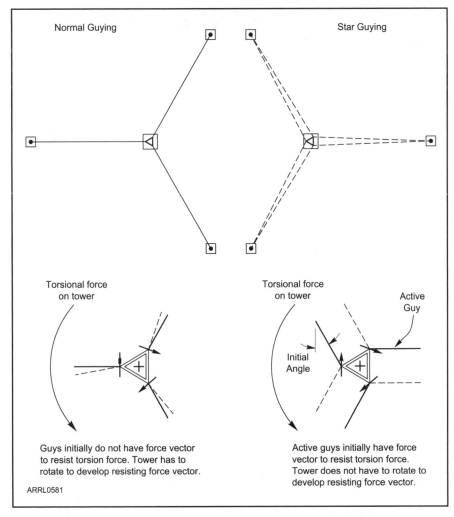

Normal Guying

Star Guying

Torsional force on tower

Torsional force on tower

Active Guy

Initial Angle

Guys initially do not have force vector to resist torsion force. Tower has to rotate to develop resisting force vector.

Active guys initially have force vector to resist torsion force. Tower does not have to rotate to develop resisting force vector.

ARRL0581

Figure 5-32 — The mechanics of star guys versus normal (single) guys.

sus conventional single-wire guys. As Hank explains:

The advantage of star guyed towers versus normal guyed towers is the significantly larger torsional force capacity the star guyed tower has going for it. You may wonder why this is important, since there are countless normally guyed towers in use. Hams, in their quest for larger, higher or more, have been known to put up stacked antenna arrays on a single guyed tower. Even more amazing is that they then want them to turn either independently or in tandem, in order to cover a wide range of azimuths. In doing so, they subject their towers to forces other than weight and wind loading. The starting and stopping of the rotating mechanisms, antenna attachment eccentricity and the use of side-mounted rotational supports all impart torsional forces into the tower. These torsional forces cause the twisting of the tower about its vertical axis.

Referring to **Figure 5-32**, you can see that the normal guyed tower has guys spaced at 120 degree intervals in plan. The guys are connected from the tower section apex to a guy anchor that is some radial distance out from the center of the tower. The star guyed tower also has guy anchors at 120 degree increments, but they have been shifted to 60 degrees in plan. The radial extension is between the apexes of the triangular tower. In addition, you see that there are two guys to each anchor point.

These double guys are typically used at all the vertical guy locations on the tower. At a minimum (to be the most effective), the star guys should be located at or near the application of the torsional load attachment points. These points include the rotator base attachment, side or swinging gate attachments and so forth. This allows for the shortest load path through the tower to the guys, and then down to the ground.

If you study the diagram, you will see how the increased torsional resistance is accomplished by the star guyed system. In the normal guyed system, the tower sections have to rotate in order for the guys to develop a force resisting vector to counteract the torsional force applied. Since the force is perpendicular to the guy initially, the guy cannot resist the force until some angle is developed to create the resisting vector.

The star guyed system, because of the double guy arrangement and location of the anchor points relative to the attachment point on the tower, initially has an angle between the torsional force and the guy force vector directions. This means that the tower does not have to rotate initially to develop a vector force resistance to the torsion. In fact, there are three such guys that would be active initially, and due to the symmetrical layout in plan, the guy system is effective in either direction.

If you are contemplating a large stacked array, either concentric or eccentric, with a swinging gate mechanism, or even an extremely large single Yagi (like a full sized 3-element 80 meter beam), the use of star guying will let you rest easy about the loadings on the tower. (Designing such a monster, well, that's another story!)

Thanks for sharing that explanation, Hank. Now we'll take a look at some practical issues to consider when working with guy wires.

Mechanical Issues with Star Guys

If you take a look at a commercial tower using star guys, you will notice that they typically simply "stack" the guy wires one above the other on the equalizer plates. Of course, I've seen as many as 12 guys terminating on one plate — *not* something most ham installations are going to emulate.

Recently, while rebuilding the multi-multi contest station at NR5M down in Texas, W2GD and I were faced with installing star guys on a variety of Rohn 45G and 55G towers. All were guyed using either Phillystran or Polygon rod, but more importantly, all utilized standard Rohn hardware. The options for terminating those guy ends were, of course, limited. The solution was to make up a rather simple adapter.

The solution involved welding to-

Figure 5-33 — The star guy adapter used in the NR5M station rebuild. See text for details.

Figure 5-34 — The star guy adapters in use at NR5M.

gether two steel plates, which would allow us to attach two guy lines to each of the existing equalizer plate's bolts. Once the adapters were welded together, we had them hot dip galvanized at a local plant. These adapters are shown in **Figures 5-33** and **5-34**.

Clearance Issues with Guys

Having guy wires sometimes means having to deal with them clearing things. Sometimes you have no choice but to run a guy wire across a driveway or over the footpath leading to your garden.

Another consideration is when an antenna must clear the guy wires. This becomes particularly important if you intend to stack and perhaps rotate one or more antennas lower on the tower. The elements of that lower antenna must clear the guys coming down from above, allowing suitable rotation of the antenna. You become concerned about the turning radius of each Yagi; you become more concerned about the guy wire interaction with that Yagi.

Computing the clearance or the required turning radius can be done by carefully sketching your installation out on graph paper, providing you with a scale model of your tower and beams. Or, you can figure out the clearance mathematically. I've used a simple formula for years, but one of my clients, Tom Warren, K3TW, modified the formula such that it also takes elevated guy post anchors into consideration, allowing me to lay out or design such installations much more easily. Here's the formula:

$$TR = \frac{d \times (g - a)}{g - p}$$

where
TR = turning radius at any point on the tower (measured above tower ground level) below an

obstructing guy wire

d = the distance from the tower base to the guy point (at ground level)

p = the height of the guy pole (use zero if guys terminate in the ground)

g = the guy point on the tower measured from ground level

a = height of the rotating antenna on the tower measured from the ground

Thanks Tom!

Ice — Considering It, Designing For It, Dealing With It

Almost every tower can withstand *some* ice accumulation. For many years, towers have been specified and designed without real formal consideration for ice loading. But ice can present a serious hazard for tower survival, especially if it occurs at the same time as high winds, or if the ice load is excessive. It's possible, for example, for ice accumulation to increase the size of guy wires by five or six times — a serious load, indeed! In some areas of the country, building codes require the tower design to assume a certain amount of ice loading. **Figure 5-35** shows actual ice chunks from buildup on an antenna. **Figures 5-36** through **5-38** show the power of an ice storm.

If icing can occur in your location, a more conservative design is called for — something you may only benefit from once, but that one time may mean the survival of your entire installation. Some designs consider that icing does not occur with full wind speeds, and look at tower loads with radial ice at 75% of the required wind loading. Some designs consider radial ice simultaneously with maximum wind.

But you must consider it. For example, say an 80 mph wind hits a 2 inch diameter tower leg. The wind load on the leg is about 18.5 lbs/square foot, or 3.1 pounds/foot. Add a couple inches of ice and that load increases to 9.3 pounds/foot!

It's just not the weight of the ice

Ice Load Danger: Near Tower Failure at W3LPL

Twenty four hours before the 1994 ARRL DX Phone Contest, for about 12 hours on Thursday before the contest the temperature was 31 degrees. A light mist was falling continuously, freezing on all of the antennas, towers, wires, guy cables and ropes at W3LPL. This was followed by strong 40 mph winds. When I arrived home around 6 PM, it was very apparent that three of the 200-foot towers were leaning over by several feet, and many of the elements on the Yagis had flipped from horizontal to vertical. The ice had started to fall as the temperature went above 32 degrees and the partially melted ice found lying on the ground was one inch thick! Only one antenna was destroyed, a 56-foot boom 20 meter beam that had accumulated a tremendous amount of ice. The boom drooped so much that the guys supporting it dropped under the then bow-shaped boom and collapsed it! Several insulators supporting wire antennas snapped, reducing the stress on the towers somewhat.

The near collapse of the towers was caused by failing to even consider the possibility of massive ice loading in selecting the height of the top set of tower guys, along with the installation of thousands of feet of wire antennas and ropes hanging off the towers — without considering the tremendous ice load they could also apply to the towers.

The fourth 200 foot heavy duty AB-105 tower was completely unaffected by the massive load because of several factors:

▪The heavy duty tower had considerable additional strength compared to the three lighter duty towers.

▪The top guys were only 10 feet from the top of the tower. The three lighter duty AB-105s had their top guys 16-18 feet from the top.

▪The ropes supporting the many wire antennas hung from this tower passed thru pulleys, then *away* from the wire antennas, *not* back in the same direction as the wire antennas (and ice) load, as on the other towers.

Lessons learned?

1) If ice is even a remote possibility, design your antennas and towers for it! I never thought one inch radial ice was even a remote possibility at my QTH. Don't forget ice loads from wire antennas attached to the towers!

2) Ice does not necessarily drop off antenna elements symmetrically! When it drops off one side first, the element is likely to flip to vertical from the tremendous weight of ice on the other side! The elements must be fastened to the boom attachment strongly enough to successfully resist this very large torque. Vertical elements could collapse the boom.

3) Guy placement should consider ice loads! Mine were much too far below the top of the tower when everything was coated in one inch of ice!

4) Ropes should attach to a pulley on the tower, then pull away from the antennas they support, not back toward the load.

5) Ropes should attach to the tower near a guy point so the guys take the ice load to the maximum extent possible.

The towers did not fall down, and we managed to repair the most critical antennas, so our score in the 1994 ARRL DX Phone Contest was only minimally affected. Then, the guying was totally redesigned and redone.

Six weeks later, a 120 mph microburst struck, destroying most of the antennas and twisting the tops of two of the towers, but that's another story!

— *Frank Donovan, W3LPL*

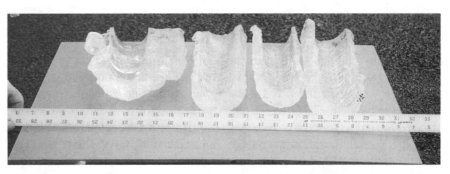

Figure 5-35 — Nearly two inches of radial ice from WN3R's mountaintop QTH in Maryland — a serious hazard for towers and antennas alike. (WN3R photo)

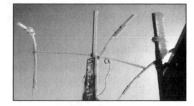

Figure 5-37 — Rime ice on antennas at N6NB. (N6NB photo)

Figure 5-38 — Pretty but deadly — the ravages of winter ice at K5CM. (K5CM photos)

alone (although that's considerable), it's how the stresses on the tower may change due to that load. A few years ago, two very tall TV towers in Raleigh, North Carolina, came down. These were relatively new, very strong, well designed and guyed 2000-foot towers. Extreme icing conditions, which lasted throughout the day and night, met the low winter sun the next day, melting enough ice off only one side of the tower so that a leg buckled on the opposite side. Fortunately, someone called the owners of the second tower, and personnel were warned and evacuated. No one was injured later that day when the second tower came down for the exact same reason. These conditions were abnormal, certainly; one cannot reasonably design for them.

Working on the tower in conditions like these is virtually impossible. It's obviously more dangerous, work takes longer, it's more expensive and so forth. So it obviously makes good sense to try planning for ice and trying to have guying and other tower-attached hardware situated and oriented properly to mitigate problems.

CHAPTER 6

Installing Tower Accessories

Once the tower is installed, there are some accessories to consider in addition to your antennas. This chapter covers rotators, thrust bearings and remote antenna switchboxes.

Rotators

A quick search on the Internet for "rotators" or "rotator repair" and you're suddenly thrust into the world of medicine — rotator cuffs and shoulder injuries and operating room scenarios of a serious and scary nature. You really just wanted to find out how to turn that beam antenna up on your tower....

The history of directional Amateur Radio antennas is a fascinating field within our hobby's general history. Early versions included the tried and true "Armstrong" method — wherein you simply turned the antenna by hand (requiring a strong arm...). Old *QST* magazines

show ham shacks with a steering wheel or something similar coming down through the ceiling — connected to the rotary antenna atop the roof.

Later, after World War II and the introduction of surplus electronics and equipment, the ubiquitous *prop pitch* motor became a *de facto* antenna rotator standard, along with the ever-popular *selsyn* synchro motors for indicating antenna direction. Prop pitch motors were originally used to change the pitch of airplane propellers, and selsyns were used in gun turrets and other places where direction or motion indicators were needed in wartime. **Figure 6-1** shows a prop pitch ad from January 1948 *QST*.

With its high ratio gearbox (the "small" prop pitch gearbox is 9576:1), which will allow it to turn practically anything, the prop pitch rotator still serves some users well today, particularly those with full-size 40 meter Yagis. Parts are becoming harder to find, and more importantly, so are people who know their way around these mechanical engineering marvels. The

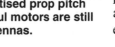

Here's the **ROTATOR** for your **BEAM!**

★ Runs on 24 to 33 volts AC or DC (4 amp. transformer will do)
★ Reversible—only three wires required.
★ 7000 to 1 Gear Reduction stops free swing.
★ Approx. ¾ RPM.
★ Powerful ¼ H.P. motor, rugged precision gear train, and sturdy thrust bearing— will support and turn even a heavy dual beam.

Used on aircraft to control pitch of propeller blades, these dependable power units are excellent beam rotators (see pages 22, 23, 29, NOV. *QST*). Used, but in perfect tested working condition, with instruction sheet............ **$12.95**
(*Mail orders add $1.25 for packing*)

Figure 6-1 — Harrison Radio Corp advertised prop pitch motors in post-WWII *QST*. These powerful motors are still coveted today for turning very large antennas.

Figure 6-2 — The original Ham-M rotator as advertised in February 1958 *QST*.

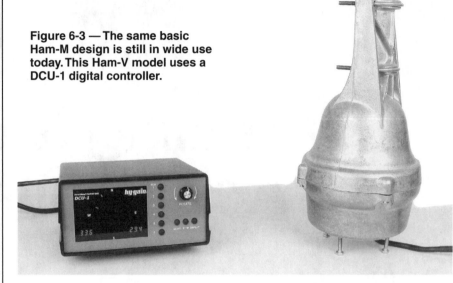

Figure 6-3 — The same basic Ham-M design is still in wide use today. This Ham-V model uses a DCU-1 digital controller.

Electronics (CDE) began to manufacture and market the bell type rotator. Early TR-2 and TR-4 model rotators were fairly light duty and often called "clacker boxes" because of the clicking noise the control box made as the rotator worked its way around to the desired direction.

Introduced in late 1957, the CDR Ham-M rotator featured a stronger case, heavier gears, a wedge brake to hold the antenna in place when not rotating, and an improved indicating system using a D'Arsonval meter movement as a direction indicator (**Figure 6-2**). Over the years, the Ham-M was updated and improved, resulting in the Ham-II, Ham-III and Ham-IV as well as a heavier duty version called the Tailtwister or T2X. CDE sold the rotator line to Telex/ Hy-Gain in 1981, and it was sold to MFJ in 1999.

Currently manufactured by MFJ under the Hy-Gain brand, the basic Ham-M design is still with us today as the Ham-IV and T2X models. It's available with the standard analog meter control box or with a DCU-1 digital control box as shown in **Figure 6-3**. It's probably the most popular and successful rotator ever produced and can certainly serve as a standard of comparison, regardless of your personal feelings about how well the design works. And it's stood the test of time — being the "model" for other manufacturers

Figure 6-4 — The AlfaSpid rotator and control box.

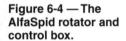

prop pitch motor isn't a plug-and-play ham rotator solution. It requires users to be part designer, part machinist and part electrician in order to realize the true potential of this powerhouse.

When television became widely popular in the 1950s, outdoor TV "aerials" appeared on the roof of seemingly every suburban home, along with the need to rotate that antenna and point it toward the TV station's signal. To meet this need, Cornell-Dubilier

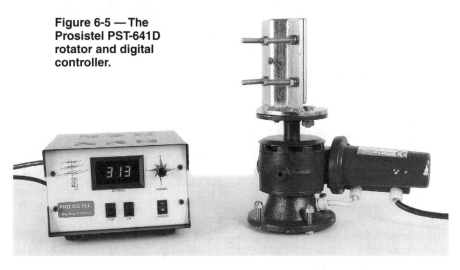

Figure 6-5 — The Prosistel PST-641D rotator and digital controller.

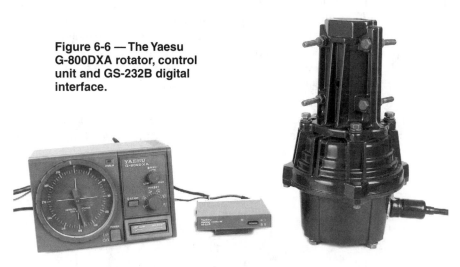

Figure 6-6 — The Yaesu G-800DXA rotator, control unit and GS-232B digital interface.

to follow, albeit with fancier braking, or improved indicator systems and the like, yet while following the original, basic CDE design.

Figures 6-4 to **6-6** show some of the other rotators available to hams. These units were reviewed in September 2005 *QST*.

Rotator Considerations

Let's go through some rotator basics, hopefully answering some questions, providing some guidelines, and ensuring success in turning *your* antenna system.

Today's amateur has a relatively limited number of choices when it comes to rotators. Most tower manufacturers provide pre-drilled mounting plates for the choices that are available. Rotator plates may be welded in place, or they may be secured with clamps that allow some choice in placement. In some Rohn G series towers, for example, the rotator shelf fits loosely inside a section, resting on Z-bracing and held in place by U-bolts. This ability to simply bolt things together can be problematic, however.

One potential problem is that most hams simply do not take the centering of their mast into consideration. The rotator mounting plate must be centered (level and plumb) within the tower structure itself. For most hams, this is something of a guessing game, compounded by the need to tighten U-bolts equally around tubular tower legs or otherwise ensure the plate is perpendicular to the mast's center.

On small tower sections such as Rohn 25, your own weight on one side of the tower is often enough to offset any attempt at eyeballing the accessory shelf into some semblance of being level (even using a small level on the shelf). This is a situation where experience counts — knowing when the shelf is level or no further adjustment is needed. There are some tricks you can use to assure that the plate is perpendicular to the mast, such as picking a secondary object in the background against which you line up the shelf. Or you can try using a small level to ensure that the plate is indeed parallel to the horizon.

Many hams do not take into consideration the actual wind load presented to the rotator by their mast, antennas and feed lines. I once encountered an 80 meter beam with a Tailtwister rotator under it, for example. It was freewheeling on 60 feet of Rohn 25G. I asked the client how many times he'd turned it. "Once," he replied. "It took an awful long time to get started, then it came around just fine, and when it got to where I wanted, I released the brake. And the beam just kept on turning."

This incident presents a good example of the ability of gearing to provide a mechanical advantage, and also a good example of the requirement of a modern rotator to hold an antenna in place once it's gotten where you want it to go. That holding power is quite often more critical than turning power.

Testing and Installation

Another aspect of working with rotators that I've encountered is the lack of testing everything beforehand! You should make a point of hooking up the rotator to the control box, turning the rotator through a complete set of clockwise (CW) and counter-clockwise (CCW) rotations. Watch for sluggishness or erratic meter movements and the like. Of course there shouldn't be any, but it's better to encounter them now, on the ground, than at height. Leave the rotator oriented *North*! (Yes,

Virginia, we're going to talk about "finding true North," in just a little bit.)

Once the rotator is oriented correctly (to North), it's time to think about mounting it inside the tower. Usually, this won't be a problem — the tower manufacturer can provide a suitable (pre-drilled) accessory shelf on which you mount the rotator, simply bolting it in place. But…be careful. Use only the supplied bolts, or if you wish to change to stainless hardware, make sure you use only the exact same size bolts! The length is often critical. Use too long a fastener, and you can damage parts inside the housing.

When you're certain the plate or shelf is level, you can slide the rotator into position. On most towers, this won't be (or shouldn't be) a problem. With Rohn 25G, the accessory shelf can really only go at one place — at a section joint. Otherwise, the Z-bracing rods get in the way. So, of course, your mast length must be considered carefully beforehand. (Rohn 25G top sections include a spot near the top without Z-bracing for the accessory shelf. This places the rotator only a couple feet below the tower top, which may not be the best solution.)

In the previous section, I mentioned the need to center the mast. With Ham-M type rotators there is a special consideration that hams often overlook. For years, the CDE/Telex/ Hy-Gain rotator manuals cautioned owners that the geometry of the mast and rotator housing accommodates *only* a $2\frac{1}{16}$-inch mast. For the typical 2-inch OD mast, shim stock ($\frac{1}{32}$-inch) is required for a perfect fit. If you are using a 2-inch mast, and you wish to shim that to the correct dimension, it's relatively easy to insert the shim stock now — before you tighten down the U-clamps in the rotator. (I like to use stainless steel material, available from McMaster-Carr.) Insert the shim stock right at the U-clamp joints, such that it's snugged up securely by the clamps as you tighten them. Don't just drop it into place and hope for the best. The idea is to "expand" the 2-inch mast diameter ever so slightly for the proper fit. **Figure 6-7** shows a Ham-IV rotator

being installed in Rohn 25G.

Many rotator designs utilize a U-clamp to provide the holding power for the mast and antenna load, and many of these clamps use only $\frac{1}{4}$ inch diameter hardware. For serious loads, I prefer something a little larger and stronger, at least $\frac{5}{16}$ inch, and I'll use the larger size if possible. I've even drilled out some rotator cases to allow me to use bigger U-bolts.

In any case, especially if the hardware is stainless steel, don't forget to use anti-seize lubricant on the threads before you begin tightening. And make sure you tighten these U-bolts as evenly as you can. And it's even a good idea to tighten them today, and then come back and re-tighten them again a couple of days later. I realize this is often impractical (if not impossible), so a quick recheck is often the very last thing I'll do at the end of the day — before I climb down. Believe me, these bolts do have a way of requiring this kind of attention to detail.

Keeping Things Lined Up

Regarding "pinning the mast" (drilling a hole and bolting the mast to the rotator housing or clamps), I have always subscribed to the notion that it's much easier to climb up and realign an antenna or mast than it is to be climbing the tower and taking broken things down. If the mast can move inside the rotator clamp, things usually will not break. I have clients who believe in (and insist on) pinning the mast, and the results have been almost exactly 50/50 when there is a problem. Half the time we merely had to climb up and readjust things. The other times, we found ourselves replacing a broken rotator (and often going up in size, to a larger rotator). I always recommend we not pin the mast, which can seem like I'm trying to guarantee future work for myself. That's not the case. I simply think it makes more sense not to force the rotator "guts" to work quite so hard.

A unique and well-designed rotator accessory is the Slipp-Nott, available

Figure 6-7 — The accessory shelf in a Rohn 25G top section is pre-drilled to accept a Ham-IV rotator. Note the terminal strip for rotator connections. New stainless-steel machine screws have been installed to avoid problems from hardware rusting over time.

from Tennadyne. This clamp, which provides considerably more surface area than the usual muffler clamp, works wonders at holding big beams in place. I've installed several of them over the years, and always found them to provide the holding power they claim.

Yaesu rotators are unique in this regard — they provide the bolt and directions on how to pin the mast. I've always believed this to be a direct result of their using a fragile mast-to-rotator clamp. It's a pot metal material that's easily broken if you torque down on the bolts too hard or tighten them in an uneven fashion. Tighten the clamp bolts snugly, drill and insert the mast bolt, and you're done. Should the antenna move out of alignment, Yaesu rotators can be realigned from inside the shack, without climbing. (And by the way, be prepared — Yaesu rotators use *metric* hardware. The very first time I installed one, I found myself at 120 feet, with wrenches that wouldn't fit anything. Nor did the client have metric tools. So he ran to Sears while I hung out up top and enjoyed watching the sailboats in Annapolis Bay!)

Bigger Antennas = Bigger Rotators

As we move up in size, with bigger towers, bigger beams and so forth, the turning requirements go up as well. For large antennas (long boom HF Yagis, 40 meter beams and even 80 meter beams), reliably turning them can become problematic, even downright troublesome, as the rotator choices are more limited. With proper gearing, most any small motor could be made to turn such antennas, but holding them in place, especially through any sort of wind, is where the problems arise. That's a tough task; indeed, it's an especially tough task. The torque requirements can become quite high very quickly. What's one to do?

Usually, the first choice will be an orbital ring rotator. The choices here basically come down to two. Those choices are the TIC ring rotator, or the KØXG orbiting ring. I've installed about a dozen TIC rings over the years.

They represent good value. Tim Duffy, K3LR, has a number of TIC rings at his large multi-tower contest station in western Pennsylvania and speaks highly of them. I believe it's fair to say that John Crovelli, W2GD, and I have more experience with the KØXG orbital ring rotators than any other USA installers. These massive rotators are truly big, heavy and simply superb at turning large antennas. They are used extensively to rotate stacked OWA antennas at NR5M (**Figure 6-8**).

Or, you *could* build your own rotator. The details on how to do just that were outlined very well in *ham radio* magazine way back in June of 1986. In "Turning that Big Array," Victor Mozarowski, VE3AIA, presented a design outline that was remarkably

similar to a commercial rotator sold by Hy-Gain, their model 3501. The 3501 was, in addition, very similar to the R9100 rotator once sold by the short-lived Advanced Radio Devices company. Most ARD products were well ahead of their time, and very rugged and well made. The R9100, for example, boasted 10,000 inch lbs of rotational torque, had a braking torque of 24,000 inch lbs, supported 2000 lbs, accepted masts from 2 to 3.5 inches, and weighed in at 230 pounds.

Or, you could rely on something like the once-popular Telrex rotator that sometimes shows up on the used market. But mounting the Telrex rotators (they go outside the tower, and move the mast through a chain drive) can be a problem.

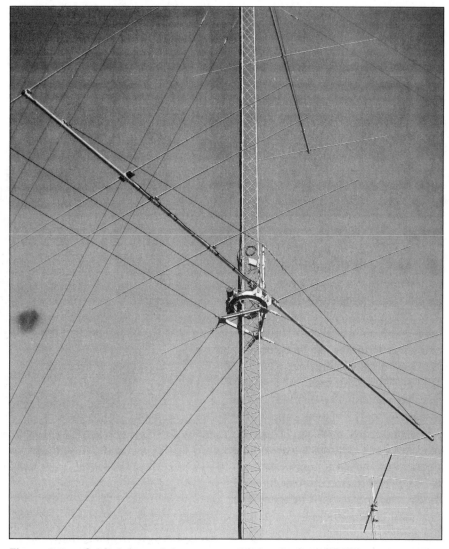

Figure 6-8 — Orbital ring rotators are used extensively at NR5M's super station to rotate long Yagis.

Or, you could utilize another old, but heavy-duty workhorse — the prop pitch motor. But as I've said, finding someone who truly knows and understands the intricacies of these mechanical marvels is getting harder and harder. Kurt Andress, K7NV, offers a rebuilding service (as of this writing) that's impossible to beat on your own; Kurt probably has more experience than anyone currently working with these beauties. I highly recommend his services.

When you reach the summit — the world of big towers and big beams — you will probably have enough knowledgeable contacts to allow you to build or buy a suitable solution to the problem of turning that big array. Check out the massive OH8X rotator gearbox in **Figure 6-9**.

Troubleshooting

When we speak about turning beams, we're usually talking about a problem — something has *stopped* turning! Ham radio antenna rotators are one of those items we typically take for granted, never giving them much thought or consideration.

Let's spend some time talking about rotators in general, and then move on to more specific areas and aspects. Not only will this allow us to diagnose potential problems more easily and quickly (perhaps without climbing), but it will also provide a better understanding of what these simple mechanical marvels should do.

Most rotator systems have three major parts: a control unit at the station end, a motor on the tower and a control cable to connect the two.

The control unit contains a power supply, a switching system and a direction indicator. The power supply usually works from commercial power, of course, and contains a low voltage dc supply. The switching system allows you to turn clockwise or counterclockwise, and to stay stopped. The direction indicator shows you where the antenna is pointed. Troubleshooting such a system is not that difficult.

If we take a look at some simplified

Table 6-1

For those using the CDE-Hy-Gain rotators, it's wise not only to log your readings in the Station Notebook, but to compare them to the typical measurements:

Ham-M Series 1 & 2

Between terminals:

1&2	0.75 ohms	Brake Solenoid
1&3	2.5 ohms	Motor Winding
1&4	2.5 ohms	Motor Winding
1&5	2.5 ohms	Motor Winding
1&6	2.5 ohms	Motor Winding
3&4	5.0 ohms	Whole Motor
3&5	Short	
4&6	Short	
3&7	500 ohms	Indicator Pot
3&8	0-500 ohms	Pot (end to wiper arm)
8&7	0-500 ohms	Pot (other end to wiper)

(Wiper is #8 for this series)

Ham-M Series 3 to 5, Ham II, III, IV, & V

Between terminals:

1&2	0.75 ohms	Brake Solenoid
1&8	2.5 ohms	Motor Winding
1&4	2.5 ohms	Motor Winding
1&5	2.5 ohms	Motor Winding
1&6	2.5 ohms	Motor Winding
8&4	5.0 ohms	Whole Motor
8&5	Short	
4&6	Short	
3&7	500 ohms	Indicator Pot
3&1	0-500 ohms	Pot (end to wiper arm)
1&7	0-500 ohms	Pot (other end to wiper)

Figure 6-9 — Sometimes there's simply no substitute for size. Martti Laine, OH2BH, stands next to the gearbox for the massive Radio Arcala rotating tower shown in Chapter 1. (OH2BH photo)

schematics, we can quickly see that there's nothing overly complicated. We should be able to determine what's wrong using some simple voltage and resistance measurements. And that's exactly the case. The key is to know where to start.

If you subscribe to the notion of keeping a station notebook, then you have an excellent starting point. You simply compare the resistance and voltage measurements on the control cable (disconnected from the control box) to those you recorded when you initially installed the system (and when it was working correctly). Lower-than-normal voltages are often signs of corrosion related problems. Resistance measurements that are suspect are also often connected to terminals or screws or joints that have suffered from the ravages of Mother Nature. **Table 6-1** shows some sample readings for HAM series rotators.

Some Rotator Hints & Kinks

While many of us unplug our station from the ac line and disconnect antenna feed lines when thunderstorms approach, it's easy to forget about the rotator control lines. Consequences can include damage to the control box or other equipment in the station. A common problem is that the Zener diode in Hy-Gain controllers can fail from a voltage spike (proving that fuses alone are not suited for such protection).

Other issues can crop up, as well. I had a client whose ⅛ amp fuse inside the control box was actually intermittent! Apparently, the weld inside the fuse's glass case had been damaged. Needless to say, this one took some time to track down, as just when we thought the unit was dead, and we'd move the control box to look for something, it would come back to life!

Connections and Connectors

While on the subject of the Hy-Gain bell rotators, some remarks on the

terminal strip or the newer pin-plug connectors are in order. I still like the terminal strip, myself, mostly because of all my years of accumulated experience. For weatherproofing purposes, I've always simply removed the terminal strip screws and replaced them with suitably sized stainless steel ones. For some folks, this proves troublesome because the original screws are "captive thread" design. That means they only screw out so far, then stop. You simply keep turning the screw and you can easily force them out without damaging anything. Then it's a simple matter to replace them with the same size panhead stainless-steel screws. I've never suffered another failure using this simple technique.

I *do not* recommend covering the screws or terminal strip with silicone seal, CoaxSeal, or anything else. I've encountered this quick fix on various jobs over the years, and in every instance, it was more of a nuisance than a solution.

I've also always enjoyed using a short pigtail (utilizing a suitable plug/jack), which allows me to troubleshoot the rotator atop the tower without bending my head half upside down in order to access the terminal strip. Ordinary eight-pin trailer light plugs and jacks serve this purpose quite well, in my experience. A number of vendors, such as The Wireman, carry them.

The new pin plugs seem to be fine, but I admit the tiny pins are not my favorite things to install, especially if you're at the end of a rather long run of large cable. Then you're often installing a smaller-sized wire pigtail anyway. Or you're carefully clipping out individual wires from the stranded cable. Neither task is fun.

With Yaesu rotators, the connecting plug (which mounts near the bottom, on one *side* of the rotator case) sticks out a considerable distance. On some crank up installations, this *will* be a problem, as lowering the top section all the way down will cause the Z-bracing to break the plug off, requiring you to buy a $50 replacement. Other than not cranking the tower all the way down, I've not found a solution to this one.

Usually, rotators do not pose problems, but when they do, it's quite often an easy troubleshooting situation. Binding issues usually relate to something being out of plumb or not level. Intermittent issues usually relate to improper connections, or to rusty or corroded connections. Direction indicator issues are almost always related to the potentiometer or the little magnet mounted inside the housing or on the motor shaft that send position information to the control unit.

Plan for Maintenance

If your installation has been planned and then carried out correctly, you should be able to remove the rotator, take it to ground level and repair it, all without creating a potential hazard atop your tower. Planning, of course, is the operative word in the previous sentence. Planning for maintenance is always a good idea, as electro-mechanical things come with an unwritten guarantee: "It's not a question of *if* something will break or fail; it's a question of *when*."

So, you should have some system planned, built, and on hand, which will allow you to do exactly that — remove the rotator without problems *when* it fails.

Cutting Cable Costs

If you've priced any of the commercial rotator cable lately, you know it can be expensive. That's especially true for the heavier gauges, which you need if your run is long. One alternative is to use Romex electrical cable — the UF or outdoor kind — instead of that expensive rotator cable.

Two runs of taped-together Romex will provide you with six wires. Then, you simply mount the starting capacitor *at the rotator* (where it does more good anyhow, rather than being down in your shack, inside the control box), and voila, this six-conductor cable works perfectly. You will need to pay particular attention to the wiring, since the colors will be limited to black, white, and a bare ground lead on each run, but that's easy enough.

Finding True North

This is necessary because north, as indicated by a compass, is not where we want to point our beams, of course.

The simplest method, or old-fashioned way, is not to use a computer or online resources. You will need to know your *local sun noon time*. That's easily done by looking up your sunrise/sunset times (use an almanac), and if the sun is bright enough, the shadow created by a vertical stake or pole at local sun noon time will lie true north.

Here's another method, not requiring local sun noon timing. Using that plumbed stake in the ground, mark the end of the stake's shadow in the morning. Later that day, mark the end of the shadow once again. Draw a line through these two points, which will be east-west. A line drawn at a right angle to this line will, of course, be north-south.

If you're in the northern hemisphere, another trick is to line up the North Star — Polaris, that bright star at the end of the Little Dipper — so you are looking at it through your tower. Mark the spot. You are standing due south of your tower. A line drawn from that spot through the tower will point true north.

Of course, if this all sounds too silly and complicated, you can look it up, or use any number of Web sites to provide the required data and information. One good resource is **www.ngdc.noaa. gov/seg/geomag/declination.shtml**. If you haven't got a local paper any longer, you can use this site to help you find or determine your local sun noon: **www.weatherunderground.com**.

Getting Your Bearing on Bearings

If you stop and consider how many things in our lives turn or rotate, a new appreciation for the humble bearing will likely evolve. Hams, of course, are concerned with turning antennas. Bearings can make such a task easier, more reliable, more precise or less limited by size.

One very useful tower accessory is the *thrust bearing*. A thrust bearing mounts at the top of the tower, usu-

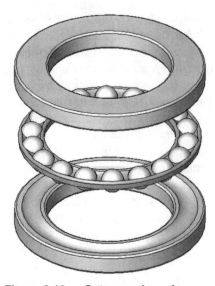

Figure 6-10 — Cutaway view of a typical thrust bearing showing the ball bearings and races. In use, the top bearing race connects to the mast and the bottom race connects to the tower.

ally bolted to a flat plate at the top of the top section or on a shelf inside the tower near the top. The mast passes through the center of the thrust bearing and extends downward to the rotator. The thrust bearing clamps to the mast, centering it in the tower and holding the weight. Without a thrust bearing, all the weight of the mast and antennas would rest on the rotator housing. That may or may not be an issue, depending on the size of the antenna load and the rotator ratings.

Bearings work by using either hard *balls* or *rollers* that move or turn inside smooth inner and outer surfaces (called *races*). See **Figure 6-10**. These balls or rollers support the load, allowing objects to move smoothly. Ball bearings typically deal with two types of load — radial and thrust — usually one at a time, or a combination of both. The typical tower thrust bearing, as you might imagine, deals with both forces, but we are mostly concerned with the vertical load that the mast and antenna(s) place on the rotator.

Mention bearings to any active station builder, and you'll get an opinion or two and perhaps even a horror story. This is one topic that's sure to draw a variety of reactions, depending on who

Figure 6-11 — The Rohn TB series thrust bearings are probably the most popular (and misunderstood) bearing in ham radio use.

you ask. Let's try to clear things up, just a bit, and answer some common questions.

Grease? No Grease?

Should I grease my thrust bearing or not? The typical Rohn thrust bearings designed specifically for tower use (model TB-3 and TB-4) are designed to run dry — without lubrication. (See **Figure 6-11**.) Their turning speed is so slow that no heat will be generated as the bearing moves. Remember, heat is what lubricants take away.

Atop our towers, let's consider some basic physics. When a ball in a thrust bearing is motionless, the load is distributed symmetrically on the ball and the race. When a tangential load is applied, causing the ball to roll, the material in the race will "bulge" in front of the ball, while "flattening out" behind it. Since not enough heat is generated from sliding friction in a typical thrust bearing, metal pickup or welding does not occur. But, the race *can* show evidence of wear. Because it's made from a softer metal than the ball bearings, the race will deform first — sometimes to the point of seizure. Lubrication will not solve that problem. Another pitfall: If the grease holds debris, it could damage the race.

You can encounter problems from large loads on thrust bearings that have set idle for extended periods. The harder steel balls will simply move the softer aluminum of the race as they rock back and forth in the wind. Over time, the bearing develops a "bump" or irregularity in the race, often enough so that when it does turn or move, the bearing will bind, causing further problems.

Multiple Bearings?

Another common question: *What about using two bearings?* While it's a good way to allow the easy removal of the rotator (the mast/antennas remain supported as you take out the rotator), alignment is critical.

Mast alignment inside the tower is one of the more important measurements we hams face, yet nearly everyone takes it for granted. Mostly, I believe, that's because there is simply so much room to adjust things. For instance, in 20 years of tower work I have only encountered one client who's asked about shimming the Ham-IV/T2X rotator mast clamp out to the specified 2.062 inches as described earlier in this chapter, even though it's clearly spelled out in the manual. Everyone assumes there won't be a problem using a typical 2.0 inch OD mast. Usually they're right, or lucky, or both and it's not a problem.

Adding a second thrust bearing magnifies the issue of even slight misalignment, which can cause drastic load concentrations. So, those two thrust bearings must be kept plumb with the mast. I usually indicate that two thrust bearings means alignment is twice as critical to my clients.

Carrying the Weight?

One further common question: *Should I put weight on the rotator, or use the thrust bearing to hold the weight?* My experience indicates it makes more sense to put the weight on the *rotator*, instead of using the TB-3 or TB-4 thrust bearing to carry the load.

Most modern rotators are designed to carry weight loads that exceed the combined weight of a typical mast and antenna(s). The bearing works best at dealing with side-to-side forces. This doesn't mean I won't use the bearing to support the mast when I remove a rotator, but for normal day-to-day operation I prefer to have the rotator carrying the load. This also seems to help mitigate that metal migration problem sometimes experienced with very large arrays (such as at large contest stations), where the load sets stationary for extended periods.

In the planning stages, it's a good idea to add up the weight of your mast and antennas and check the rotator ratings.

Other Choices?

Next, that inevitable query: *What about using something other than Rohn bearings?* And the immediate answer is, of course, it's certainly possible. Such bearings usually only have one, or at most, two, setscrews — not three evenly spaced ones as used in the Rohn design. The major limitation of most bearings is they are intended for shop use — indoors. They will rust quickly when used outside in the weather. Yet, I continue to see them used by clients and supplied by various vendors and tower manufacturers.

Here are some preventative measures I've used successfully in protecting shop bearings:

■ Paint the entire bearing — flange and bearing surfaces — with Rustoleum Rusty Metal Primer. Use a minimum of two coats.

■ Replace the Allen head set screws with hex-head bolts. These will be fine thread (NF) bolts and will probably require some searching. But replace them now, before installing the bearing, or you'll be drilling out set screws one day, atop your tower, certainly something that's never a fun task!

■ Be prepared to climb and grease such bearings, usually twice a year. Remember that water can migrate into the grease and freeze. Remember, too, that this same grease can hold other debris or contaminants, some of which may lead to premature failure. Keep such bearings clean!

■ Try to cover the bearing with a suitable boot or cover of some kind. Each installation will require something different, but the business end of a plumber's friend, some vent stack covers, or other building parts can and do work. I usually use a large remnant from a truck tire inner tube, cut to shape and held to the mast by a hose clamp.

Bushings

You may not need or want a real bearing (required for supporting vertical loads), but just a simple bushing. This works especially well if you re-

Figure 6-12 — A 3-inch bearing/bushing made from Delrin stock and installed on K4VV's Pirod tower. (K4VV photo)

member that you can reduce bending moment forces if your mast extends *below* the top of your tower about as far as it extends above the top.

Most modern rotators can support typical heavy mast and antenna loads easily enough. A bushing can be used to properly align the mast and provide an appropriate surface for the mast to rub against to prevent wear as it contacts the tower top section. **Figure 6-12** shows a bushing made at K4VV's QTH with some Delrin stock and installed on his Pirod tower, supporting his 40 meter beam.

Further Reading

In true ham spirit, when talking about towers and thrust bearings I'm always reminded of an article by Fred Hopengarten, K1VR. Published on the Yankee Clipper Contest Club Web page (**www.yccc.org**) and listed under "Articles," it's called, simply, *FAQ: Thrust Bearings.* This document is filled with neat, often witty, but always useful facts and ideas. It ends with a step-by-step description of how to refurbish a Rohn thrust bearing on your own. (Another good resource comes from Jim Idelson, K1IR, and was also published as "Refurbishing Your Rohn TB-3 Thrust Bearing" in the Sep/Oct 1999 *National Contest Journal*.)

Turning Everything At Once

Turning the entire tower has become more popular within the past few years — as stations get more advanced (or complicated, if you like). There are some definite advantages.

You can install bearings and rotating rings on a lattice-style tower, or you can take the rotating monopole approach if guy wires are an issue for your installation. In either case, you'll be turning everything at once!

As stacked arrays increase in popularity and use, turning them can become somewhat troublesome. A rotating tower moves all the antennas together, at once, in echelon. While an avid contest station might not want to turn everything at one time (antennas pointing in different directions can be an advantage), a DXer might want to maximize forward gain at all times. A VHF/UHF station could also benefit from turning everything at once to take advantage of stacked antennas or to line up antennas for various bands on a distant station.

To more fully utilize the benefits of a rotating tower system, a single tower can be separated. Typically, the lower half remains stationary and the top half rotates — providing some variety in directions. Assembly of rotating tower hardware is usually no more difficult than simply building the tower itself. Most rotating tower systems are built on nothing less than Rohn 45G and 55G, both of which are heavy-duty hardware. So, be prepared — it's heavy stuff!

While rotating the entire tower is a specialized approach to rotating your antennas, there are some commercial options or accessories that make the process easier. I've installed both Rotating Tower Systems (RTS) and K0XG hardware. Both are extremely well made and heavily galvanized. Despite the mass and weight, they are easy to install and maintain. RTS rings must be installed as the tower is being constructed (at section joints); K0XG's rings can be put on the tower at any point. Bolting the antennas on to the tower (side-mounting) can also be simplified if you utilize some of the K0XG mounting hardware, for example.

Having the rotator located at the bottom of the tower allows for easy servicing, even easy replacement, should that ever be necessary, another advantage of a rotating tower setup. And you'll be amazed when you find you can turn the entire tower by hand prior to installing the beams and rotator!

Stacking Antennas

If you have a heavy enough tower, you may wish to utilize some of the space available and stack antennas lower on the tower. There are a number of switching/matching devices on the market that will allow you to feed antennas separately or in combination. You can keep the lower ones all pointed in one, fixed direction, or you can aim different antennas toward various population dense areas. A popular approach is to rotate the top antenna(s) and point the lower antenna(s) toward an area of interest (say Europe or Japan).

Mentioned previously as a solution for large Yagis, orbital ring rotators are a great solution for rotating lower antennas of any kind around the outside of the tower. They'll handle Yagis large and small; the main concern is clearing guy wires.

Another approach to rotating lower antennas is a sidearm mount. The swinging gate sidearm enjoyed some popularity in the 1970s/80s, and it is still a viable approach, especially for the clever homebrewer. The main drawback is that the mount or antenna boom will hit the tower at each end of rotation, so you're limited to about a 300° sweep depending on design.

WB0W sells a popular sidearm mount. Array Solutions markets a clever swinging gate sidearm, and Max-Gain Systems sells the IDC Technology Sidewinder swinging gate. IIX Equipment offers a swinging sidearm mount as well. So, there are options if you have the room on your tower and wish to benefit from stacks of antennas.

Remote Antenna Switches

Once you have a tower, it's a natural inclination to utilize it not only for that HF rotary beam (or beams), but also as a high support for wire antennas and perhaps even a VHF or UHF antenna (or two).

So it's natural, as well, that you then find yourself confronted with the need to run several coaxial cables to feed your growing crop of antennas. Rather than run an individual feed line to each antenna, it makes good economic sense to run one *very good* feed line (large diameter flexible coaxial cable, or hardline, or Andrew Heliax, for example), and then remotely switch the various antennas in and out of connection to that feed line as needed. This arrangement may make practical sense as well, compared with routing multiple feed lines to your station and considering the ease of adding or changing antennas.

If you want to go this route, that obviously means you're now in need of a switch. Remote antenna switches must be designed for outdoor use. The cover or mounting box should be weatherproof (even waterproof), and yet allow easy access to the appropriate jacks to attach feed lines going to the various antennas. If the enclosure is not weatherproof, eventually the switch will become intermittent or fail completely.

Hams are fortunate to have a rather large number of potential choices. Typical models allow the choice of four or more antennas (**Figure 6-13**). They typically use relays with ratings suitable for high power RF operation and require a multiconductor control cable. Most offer manual antenna selector switches; some are designed to take advantage of station automation and computer control. Some specialized models are designed for stacking and switching among two or more antennas for additional gain.

It's impossible to pick one manufacturer's product over another, as the

Figure 6-13 — A remote antenna switch allows a single very good quality (expensive) feed line from the station to be used with multiple antennas on the tower.

requirements and options will vary widely from station to station. Some things to consider are power and frequency ratings, control options and requirements and connector style. It's a good idea to buy a remote switch with extra positions in case you want to add antennas in the future. I've had very good results with units made by Heath and Drake (available used),

Ameritron, DX Engineering and Array Solutions, at my own station as well as at the installations of various clients. Check the ads in *QST* for these and other options.

When installing a switchbox, I usually make sure the mounting hardware is stainless steel and change it if need be. I usually try to mount the switch such that the cable access is easy when I'm standing on the tower. It's often difficult to weatherproof SO-239 connectors clustered together on the underside of a box, but make the effort. It's possible to wind some tape backward (with the sticky side out, wound carefully) around a small dowel or drill bit, and then unwind it onto the connection. This process is much harder to describe than to actually do, but it takes real practice before you can succeed with this old-time electrician's trick.

The switchbox is also a good place to utilize an under-used tool — the coax boot, a rubber or vinyl cover that slides right over the entire coax connector. This upside-down location is already partially shielded, and the boots provide just the needed measure of protection, easily and quickly. Just remember, you'll need to install boots *before* you install your connectors on the coaxial cable! I've found the boots will last four to five years before drying out and cracking, even longer if you can cover them suitably.

CHAPTER 7

Working with Cranes and Lifts

Using a crane to help erect towers and/or antennas is not as expensive or difficult as you may think. You might consider hiring a crane when the mechanical advantage is simply too great (compared to lifting big, heavy objects manually), or safety factors outweigh the extra expense, or time savings alone makes it worthwhile. Sometimes a damaged tower or antenna is simply not safe to work on and a crane is the only practical solution.

Whatever the reason, sometimes using a crane is simply the better, safer solution when putting up or taking down towers or big beams. Accordingly, some thoughts are presented here on working with cranes and lifts.

Crane Background

If you've never worked with or used a crane, it's important to understand that cranes are mainly used to hoist heavy loads and move them *short* distances. A crane is a steel boom (usually) mounted on a hinged hydraulic platform with a cable running through it on pulleys. This cable ends in

a hook and is raised or lowered by a winch attached to the platform. The boom can be raised and lowered, and the entire platform can turn, or slew.

Most cranes today are hydraulic. Most of us will use mobile cranes, which can vary in size and shape, depending on their lifting capacity or boom length. Truck-mounted cranes such as the one shown in **Figure 7-1** can be cost-effective. They have boom lengths going to 120 feet, yet are still small enough to fit into a normal backyard. Most of the time, weight limits will not be a problem for cranes, as they're designed for construction

use and ham towers or antennas are extremely lightweight in comparison. The crane is heavy, so you'll need to ensure that the equipment can reach your tower site without getting stuck in soft ground or ruining your flower beds or sprinkler heads or septic system.

Like other heavy equipment, cranes are rented by size (capability), by the hour, with higher rates for more capable equipment. Rental almost always also means "portal-to-portal," which means that you'll pay for time required to get the equipment to your location and back to the vendor's shop. Also, some states (such as Maryland) impose fees for transporting such heavy vehicles on the highway. When getting an estimate from vendors, make sure to ask about all applicable fees, taxes and other charges.

Figure 7-1 — A small truck mounted crane installs K9US's tower in one pick.

Putting a Crane to Work

The first time you use a crane on a job, it can be a sobering and frightening experience. Once you do it, you'll be hooked (pun intended) — no more heavy lifting, no more laborious rigging of tramlines or back guys, and so forth. A crane can make easy work of setting a heavy crank-up tower in place as shown in **Figure 7-2**. Guyed tower construction can benefit from use of a crane to lift long assembled sections of tower, complete with guy wires, mast and other hardware, into place as shown in **Figure 7-3**. It can also make short work of lifting a large, fully assembled Yagi into place (**Figure 7-4**).

Of course, the crane operator may not be familiar with ham towers or antennas. They're used to lifting large, heavy items, not delicate fixtures of aluminum. Plan on spending some time going over the work to be performed.

I always include some talk about our terminology with the crane operator beforehand, too. If he's suddenly confronted with someone yelling about the beam or Yagi (terms he's probably never heard before), it may be too late to save the item in question from damage. I almost always spend way more time talking about the job we're about to do than it actually takes to *do* the work. But, as usual, safety first is the key to this work, and such discussions help ensure a safe experience by getting the climber(s) and crane operator on the same page.

This is a good time to point out that the tower climber(s) direct the op-eration. They're the ones at risk, and more importantly, they are the ones who are the closest to the action. They can truly determine what's going on and what needs to be done next. It's vitally important that climbers communicate correctly with the crane operator, whether that's via radio (such as a pair of FRS handhelds) or through the typical hand signals. In either case, a brief meeting of minds is called for before starting work so everyone is in agreement on what's to be done and how the work will progress.

Rigging is always a factor to consider — how do you attach that delicate Yagi to that huge steel ball (300 to 600

Figure 7-2 — A small truck mounted crane makes easy work of installing K4KL's 89-foot crank-up tower on a prepared base. For a tower of this size and weight there are not a lot of safe options.

Figure 7-3 — K4ZA and helper at the 40-foot level await the second half of W9GE's tower. The mast, rotator and guy wires are already installed, a perfect example of the savings in time and labor a crane can offer.

Figure 7-4 — A crane lifts a fully assembled OptiBeam OB17-4 to the top of N3KS's tower. The antenna has 17 elements for 40, 20, 15 and 10 meter — the crane makes installation a lot easier. (Melanie Sirageldin photo)

pounds, typically) hanging there with a simple hook at the end of the heavy steel cable? I like to rig the load myself, and then climb. I use my usual assortment of slings and carabiners to rig antennas, using techniques described in earlier chapters for working with ropes and ground crew muscle. These tools are much smaller and lighter weight than what the crane operator is used to seeing and using, so discussing them beforehand is another good idea.

Time spent rigging and testing by lifting the antenna or tower section a literal few inches will help show everyone that your load is balanced, secure and ready to "fly" once you're atop the tower waiting for it. After you and the load leave the ground, you'll need to communicate with the operator via radio or hand signals. If you don't have a suitable radio, you must rely on hand signals. A simple Internet search will provide the half dozen most common ones (they *are* standardized). Learn and use them. Again, safety first!

Human Cargo

Sometimes, lifting an antenna into place isn't needed. What's needed is a fix or repair to an antenna already mounted and in position on the tower. The obvious option is removing the antenna and taking it to ground, effecting the repair(s), and then hauling it back into place. Another way is repairing the antenna *in the air*. If your boom-to-mast mount doesn't allow you to easily move the antenna so you can reach the damaged part from the tower, what can you do? One solution is using a crane to bring the repairman to the problem. Basically, there are two options.

First, and the simplest, from a hardware point of view, is the so-called

bosun's chair (**Figures 7-5** and **7-6**). The technical definition of a bosun's chair (sometimes spelled *boatswain's* chair) is a device used to suspend a person from a rope to perform work aloft. Often little more than a board and some rigging in the sailing world, the bosun's chair allows the climber to access the antenna while suspended from the headache ball of the crane. Working from a bosun's chair limits maneuverability. One hand is usually needed to hold oneself in place, for example — especially at real heights where the wind is often a factor, even on calm days. But for quick and simple repair or adjustment tasks, it's sometimes a workable solution.

A more serious approach to being hoisted into working position by the crane involves what's known as a man basket, crane basket or personnel platform. These rugged, steel cages are also suspended from the headache ball. There are one man and two man baskets, and if you're required to spend more than a few minutes in the air, it's the only logical choice. I once spent half a day in one, going around and around the 3-element 80 meter Yagi at

Figure 7-5 — K4ZA rides the bosun's chair to 125 feet to replace the broken reflector on one of NR5M's 15 meter Yagis. (K7ZV photo)

Figure 7-6 — K4ZA and W2GD repairing one of NR5M's 15 meter Yagis at the 165 foot level. It was faster and easier to bring K4ZA to the work than to bring the Yagi down to ground level. The bosun's chair provides relatively easy access.

Figure 7-7 — You've probably seen the Genie S-40 boom lift at construction sites. It's a great way to install or work on antennas or tower hardware.

NR5M, when we were in the final tuning stages, literally adjusting each element by inches.

Man basket rental is usually an additional expense and must be ordered with the crane when placing your order.

For certain jobs, it's the most practical answer to the question of how to solve the problem of working successfully while "hanging out" in space.

Lesser Lifts

Say you want to work on your antenna, mounted on a crank up tower that is cranked down to 20-something feet. In that case, you don't really need the height or lifting power of a crane. Consider an *aerial lift*, instead. Often referred to as *manlifts*, these aerial platforms have become increasingly popular over the years and are used at nearly every construction site these days. These lifts can be found seemingly everywhere today, at reasonable rental rates. They provide a safe platform from which you can perform a variety of tasks and jobs. They're much safer than ladders or scaffolding.

From the straight boom lift (**Figure 7-7**) to the articulating arm lift (sometimes called a knuckle boom), to the scissors lift and the personnel lift, there are a variety of options designed to get you working safely up in the air. All without climbing!

The boom lift is a straight, telescoping length of boom. Typically available at some larger rental houses up to 120 feet,

Figure 7-8 — An articulating manlift can be maneuvered in between Yagis in a stack. In this case the manlift provides a safe and convenient way to finish the antennas on K4KL's crank-up tower.

on an all-terrain, self-propelled mount, these lifts can solve a variety of problems. The articulating arm lift, which is usually in two sections, can shift around objects (getting you in between elements of a long boom Yagi, or under and around guy wires, for example), putting you exactly where you need to be to work. **Figure 7-8** shows how useful this tool can be. These booms are usually not quite as long, but again, the mounts are usually all terrain, so they can be driven virtually anywhere. Both the boom and articulating arm lifts can usually carry 500 pounds.

The scissors lift has tremendous lifting power (often up to 2500 pounds), but all of it is confined to straight up and down directions. They're usually available on interior surface (smooth) mounts only. The personnel lift is intended to lift one person straight up and down, and again, these are intended mostly for interior use. But they can be very lightweight, often wheeled into position by hand.

Most of the larger, self-propelled units are quite easy to operate, and the rental company will gladly provide you with the simple instruction needed.

LIFT SAFETY — A Few Simple Rules

Like any tool, manlifts are safe only when they are used correctly.

Falls or tipovers and electrocutions are the main cause of death in manlift accidents. So, *pay attention* to what's overhead or nearby, just as you do when working on towers. While OSHA requires commercial workers to wear a safety belt in the basket to prevent falls, as amateurs, we can opt not to do so. I've never found wearing the safety belt to be a hindrance to working quickly and easily, and recommend wearing one any time you go aloft.

Always be certain of the terrain. Never work on sloping ground. Never work on extremely wet ground. Always use outriggers, if supplied. If uncertain of the terrain, chock the wheels for safety.

Don't climb up on or lean over the guardrail of the man basket.

Don't exceed the load capacity limit (consider the weight of tools and equipment, plus yourself).

If working anywhere near vehicular traffic areas, set out cones or warning tape to keep traffic away.

Do not exceed the vertical or horizontal reach limit of the lift.

Do not move the platform with the boom extended. Always orient the lift so it's moving in the direction you're facing.

Communication is critically important. The person in the basket should always communicate with ground personnel before moving the basket.

The tow-behind lift can be moved onto your jobsite, outriggers placed and ready to haul you up in the air, all literally within just a few minutes. Battery powered, these lifts can reach up to 60 feet in working height. The only downside is you almost always need to be well versed in the fine art of backing up with a trailer in order to maneuver the lift into the proper working position.

If you need to work "up in the air," aerial lifts offer a safe, relatively inexpensive alternative to old-fashioned methods. They are often available at large equipment rental facilities.

CHAPTER 8

Getting Antennas Up in the Air

At some point, unless you have someone to do the work for you, you'll find yourself needing to get an antenna up off the ground and onto the tower. If it's something you've never done before, you may wonder exactly what the best, and safest, method is for doing the job.

There are several possibilities. Picking the right one will depend somewhat on the specifics of your installation, your experience and comfort level, the available tools and equipment, and your budget. You could use a crane as described in Chapter 7 — provided there's room in your budget, as well as your yard. You could assemble the Yagi up in the air — at the top of the tower or above the lower guy wires, perhaps. You've read or heard about *tramming* antennas into position on the tower. But what, *exactly*, are your options? How can you accomplish this often seemingly Herculean task — especially with a guyed tower? We'll briefly describe other options, but tramming is the main topic or focus for this chapter.

I suggest tramming as the main topic because I believe it's the one of the more efficient ways to move antennas up and onto your tower. Yes, there are a variety of steps and procedures required in order to accomplish the task. Yes, there are some specialized tools. Yes, there are some space requirements — you must have room to maneuver, not only the Yagi(s), but the ground crew, as well. Yet, the benefits far outweigh

all these requirements in my opinion.

As always, if you ever have any doubts or questions about rigging, about procedures, about safety issues, always ask them before life, property and peace of mind are at risk.

Tramming Antennas

Tramming is often misunderstood. Whenever someone brings up tramming on Internet discussion groups (such as Towertalk, accessible from **www.contesting.com**) it always generates a lot of traffic. **Figure 8-1** shows the general idea: The assembled antenna rides to the top of the tower on a tramline affixed to the mast. The tramline is anchored near or to the ground at an appropriate distance from the tower base, and rigged so that the antenna boom and elements avoid contact with guy wires.

First, let's get on the same page with our definitions.

▪ *Tramming* means the antenna is suspended *beneath* the tram line.
▪ *Trolleying* means the antenna rests *upon* the line.

There are considerable differences in approach and in using the two methods. I much prefer tramming, myself, although I've found myself in situations where using a trolley was the better solution. This chapter will, however, focus on the tramming method. It

seems more practical and more useful for nearly all applications.

For me, successful tramming depends upon the use of many of the climbing gadgets discussed in previous chapters — slings, carabiners, proper ropes and pulleys. Without them, I'd be lost. With them, I can accomplish my rigging safely, securely, easily and relatively quickly, too.

Balance is Key

The first thing to do when setting up a tramming job is to ensure the load (the antenna) is dynamically balanced. The center of gravity should be as near to the boom-to-mast clamp as possible. If the antenna is *not* balanced, not only will rigging be extremely difficult, but also the antenna will not want to "fly flat and level" on its upward journey. So, the first thing I try to do is balance the antenna.

The first and easiest thing to try is simply shifting the boom-to-mast clamp's location. This may not be possible on some antennas because of element locations or the clamp design.

If you cannot balance the antenna by moving the clamp, the next step is often inserting an appropriate weight inside the boom. I always try to add any weight as near to the boom truss attachment point as possible (most longer boom Yagis use a boom support truss of some kind). I use galvanized water pipe of an appropriate size

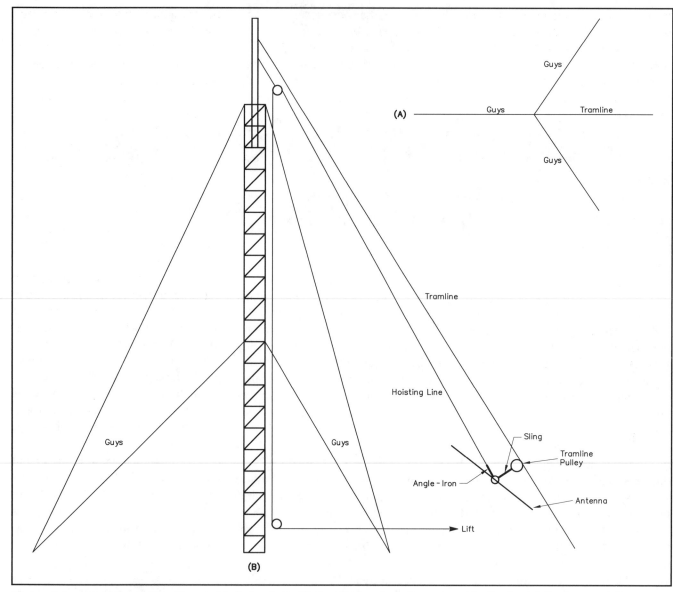

Figure 8-1 — The tram system is used to bring large Yagi antennas from the ground to the top of the tower while avoiding guy wires and other obstructions.

inserted inside the boom. For example, 2 inch galvanized water pipe weighs 3.65 pounds per foot. The pipe is usually secured with a stainless steel bolt, passed through the boom. If the weight itself is small enough, simply squirting in some expanding polyurethane foam will hold it in place. See **Figure 8-2**.

Balancing the antenna requires that you be able to lift the entire beam high enough off the ground to allow you to judge when you've reached the center of gravity. This means lifting it high enough to ensure element tips are not dragging on the ground, and so forth. To determine the amount of weight required for balance, I pull down on

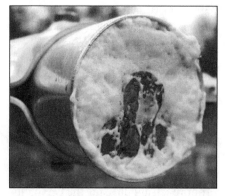

Figure 8-2 — Expanding foam insulation is used to secure a weight inside the boom of a 10 meter Yagi for dynamic balance. (K8PO photo)

the light end, using a strain gauge that registers up to 50 pounds. (*Strain gauge* is a fancy name for a scale used to weigh fish, available from sporting goods stores. See Appendix A.) Having someone sight along the boom and telling you when it's level makes this job relatively easy, of course. Then, you simply read out the weight needed to reach that balance, and insert that amount into the light end of the boom, again, as near to a truss attachment point as possible. This may require some judicious jockeying back and forth, as the truss attachment point will not be exactly at the boom end. Time and experience pay off here, and it's easy enough to get the beam into balance with a little patience.

Remember, adding the coax that runs along the boom to the antenna feed point will often upset your carefully planned balance point if it's not installed during weigh-in, but that can often be compensated for through careful rigging.

Rigging the Antenna

At this point, many folks think they are ready to begin tramming. Not quite. There's no guarantee (even if you've balanced the beam) that the antenna elements won't "roll over" or shift direction as you begin lifting (a perennial problem with KLM antennas, for example). The first time you try this, you may be dismayed to discover the beam's element tips suddenly catching in your guy wires on the way up. Or, perhaps your boom slowly slides to the right or left, wandering away from the attachment point on the mast.

Some folks will try using tag lines that are pulled as necessary to control this side-to-side movement. And, tag lines can and do work, but their removal (once the beam is in place) sometimes leaves a lot to be desired. There's always the chance a tag line may get caught or snagged during the removal process, for example. That's never fun. Here's one solution to not using tag lines.

Using slings, suspend the beam just below the tram line — right under the upside-down pulley that's resting on the tram line. Orient the elements toward the tower, of course. Again, the idea is to keep the beam level. Next, to connect the antenna to the rope you'll use to pull it up to the tower, you need to build one small additional piece of hardware often referred to as a *tiller*. It's really more of a torque arm, and it works like a lever. (Glenn, K6NA, is often credited with this idea. His remarks about tramming on the Towertalk reflector are well worth some research time!)

The torque arm is a simple piece of angle iron (or angle aluminum), about 2 feet long and ⅛ inch thick by 1.5 or 2 inches wide. Drill one end for a muffler clamp that will fit your beam's

boom; drill the other end to accept a shackle or carabiner.

Attach this torque arm on the boom as near to that center of gravity as possible, which puts it very close to the boom-to-mast plate. With the torque arm pointing toward your tower, you'll find that you can now "set" the angle of the elements themselves as they go up (relative to the tower), by simply moving the arm.

Pushing it ever so slightly lower (below the plane of the elements), for example, will cause the element tips to rise higher, and so forth. Ideally, you'll find a small degree of shift relative to the tram line will probably be about perfect, which is the effect you'll find if the arm is parallel to the elements. This arm will keep the elements away from the guys, and you'll find it provides the control you need when tramming. Too much angle of attack, and the elements can collide with those of another antenna that may already be in place on your mast!

I think you'll find this method of tramming works quite well, and it even makes a one-man tram operation possible, albeit one requiring lots of climbing. You will still have your hands full with some of today's longer-boom Yagis, pun intended!

The Tram Line

Tramming does require some room around the tower and may be difficult or impossible in wooded areas or on a tight lot. When locating the anchor point, it's advantageous to keep the angle of the load to the tower as shallow as possible. A shallower angle provides more control over that "angle of attack" of the load going up.

I always prefer to rig the top of the tram line right onto the mast itself. Another choice is to rig it to the tower, above the top set of guys, and then pull the load up into its final position, on the mast, either by hand or with a come-along if needed. Again, the climbing slings and associated hardware make short work of these tasks. They're lightweight, small, and solve many rigging issues quickly and easily.

I do not attempt to use my gin pole

for tramming unless it's with very small loads such as VHF/UHF antennas, small tribanders, or the like. The mast itself simply works much better. Back-guying the mast is a good idea if the load is large, or if the mast is aluminum, for example.

Material for the tram line itself varies with the load you're going to lift. I like ³⁄₁₆-inch EHS guy wire for big Yagis, although I've used ⅝-inch rope, too. With EHS guy wire, a Preformed Line Product Guy Grip (discussed in Chapter 5) makes an ideal attachment method at each end, of course. I also like stainless steel aircraft cable, as it's very flexible and somewhat easier to maneuver around than the EHS, but you're back to fussing with cable clamps once more to secure the ends of the tram line. Whatever you choose, make certain you are securing the tram line to something solid at the ground end.

The critical factor here is to use an appropriate pulley for whichever material you choose as your tram line. The beautiful nylon and aluminum rescue pulleys will not work on the EHS or aircraft cable, for example, only on rope. **Figures 8-3** and **8-4** show two

Figure 8-3 — A sturdy all-metal pulley used for tramming.

Figure 8-4 — A tandem pulley works well for tramming because it distributes the load more evenly on the line.

Figure 8-5 — Tramming a KLM 15 meter Yagi at NR5M. W2GD is waiting at the 105 foot level to mount the antenna to a TIC ring rotator.

pulleys suitable for tramming on wire rope or cable.

Don't set up the tramline so that you're unable to lift the load (the antenna) up and onto the tramline itself. Once the load is on the line, it's a good idea to allow it to settle, again making certain it's balanced.

Lifting the Antenna

Once the rigging is in place and the antenna balanced and secured, it's time to haul the antenna up the tramline (**Figure 8-5**). Usually that's done by hand, with that old standby — manpower — pulling carefully on the haul line. When possible, I prefer to use a capstan winch (see the sidebar) to

The Capstan Winch

If you're going to be doing any serious tower work, at some point you'll eventually encounter the capstan winch. A logical application is tramming antennas or hauling heavy hardware, such as rotating tower bearings or orbital ring rotators, up the tower. Probably no single mechanized tool is more misunderstood or misused than this simple device.

As seen in **Figure 8-A**, a capstan winch is a motor-driven spool or drum around which you wrap 3 to 4 turns of your haul rope. The tension you put on that rope (literally fingertip pressure) determines the amount of pull applied to the load. The rope never accumulates on the drum, so there's a constant rate of pull — perfect for hauling stuff up a tower.

Capstan winches are designed to use small diameter synthetic ropes — materials with low stretch characteristics. The winch itself needs to be mounted securely, in line with the tower base. It's often convenient to mount the winch either on the tower base itself or on a trailer receiver hitch on a truck. Either way, make certain the angle going upward on the tower is free and straight. If you use the truck mount, a separate sheave or pulley mounted at the tower base is required. Make certain that the haul rope goes on and comes off the capstan drum at a right angle — so that it cannot be spooled off the drum during use.

If this discussion sounds somewhat vague, and you'd like specific, detailed directions on setting up a capstan winch, please realize that this is one area in which there's no substitute for experience. For each installation, setup will require that the rigger draw upon not only common sense, but on his own, personal experience and techniques.

Lay out the haul rope so no snags are possible. The rope between the pulley at the tower base and the winch drum should never touch the ground during lifting, for instance. Don't stand so close to the drum that you (or your clothing) could possibly get snagged or tangled in the line. *Never* wrap the rope around your hand, arm or body.

Most winches utilize a "dead man" style switch for power — usually a foot pedal type. *Never* defeat such switches. If the winch is spinning, but the load isn't moving, stop *immediately* and determine why. You may simply have chosen the wrong kind of rope — one that has simply reached its elongation or stretch point. Or the load may be snagged. Stop, secure the

Figure 8-A — The capstan winch (My-Te model 300) mounted on K4ZA's truck, ready for work.

rope, and figure out what's going on. To stop during a lift, simply step off the dead man switch and hold the line. Do not simply allow the motor to turn while loosening the line — that causes friction, heating up and possibly burning the rope.

Proper care and inspection of the capstan winch rope is more critical than for other ropes found in the tower climber's toolkit. (I have a rope dedicated solely to my winch.) Avoid kinks, dirt, knots and snags in the capstan rope. Any of those could cause the drum to lose solid contact with the haul rope, making you lose control, which could be disastrous.

A capstan winch is a simple, useful tool. It's not used every day, or on every job, obviously, but it's a truly valuable tool for saving labor and time.

handle the lifting. For some really big and heavy loads, the winch represents the only logical solution.

I do not like using garden tractors, pickup trucks or other vehicles to pull the haul line. Others may view the use of mechanized help or vehicles differently, but in my opinion, they isolate and limit the control entirely too much. Quick and accurate communications are vital during the tram operation, and shouting to be heard over a tractor motor and exhaust isn't very practical. Also, these mechanized power sources will keep on tugging, regardless of conditions,

and possibly to your dismay if not stopped in time.

Keeping a hand on the haul rope, or even standing in line with it, is not a good idea if you do resort to machinery to lift the load. Should trouble develop (for example, a pulley binds or the antenna catches on guy wires) and you do *not* stop in time, the whiplash action of a breaking rope could do real damage to anyone or anything nearby.

Regardless of your setup, it's important to keep a watchful eye on the haul rope throughout the process. Its movement (or lack of movement) can provide you with clues that will pre-

vent disaster much faster than anything else. This is another reason why the climbers up on the tower can better judge what's going on at all times, simply by keeping a watchful eye on the haul rope.

If all this talk about rigging suggests that tramming is overly complicated, I've given the wrong impression. It does require some extra setup time, some small amount of experience and some specialized tools. But these potential disadvantages are far outweighed by the ease in getting your antenna(s) up in the air and safely onto your tower. Once you get used to it, or even good at it, you'll find yourself in the enviable (or dubious) position of being asked to help do it for your ham friends and their own installations. Follow these suggested guidelines and you'll indeed be successful!

Options to Tramming

With self-supporting towers, it's usually a simple matter to simply hoist the antennas straight up the tower and into place on the mast. Sometimes the process is more complicated because the tower is tapered. Or, if it's a crank-up tower, standoffs to support the coax and control lines or winches and other hardware located directly at or on the base may be in the way. A simple tool can help with these situations. It's a simple gantry type device that guarantees that the load will be held out, away from the tower, and can then be pulled straight up. Details on this tool may be found in the **Appendix A**.

Trolleying

Working with a trolley, where the antenna is suspended on the line rather than underneath, is more difficult. It requires two cables or lines, not one, which means you need more cable to begin with, along with the hardware to attach them.

Setting and maintaining *equal* tension on these lines is critical to success, too. And then, there's the actual trolley itself — something you will have to construct or make. The trolley will have to be strong enough to securely carry the antenna's weight, rigid enough not to flex itself apart as the lifting and pulling forces exert pressure on its upward journey, and simple enough to allow you to detach the beam from the trolley and attach it to the mast once at the top of the tower.

If this sounds complicated, I'd agree. All further reasons for choosing tramming!

Working Around Guy Wires

Sometimes, tramming isn't possible and another way must be found to

Figure 8-6 — K4ZA installs a TH-11 Yagi for N4IZ by corkscrewing it around the guy wires.

move an antenna up and onto a guyed tower. Sometimes, it's possible to accomplish this by using a procedure commonly called *corkscrewing*, which means the antenna is simply turned (in a corkscrew fashion) around each individual guy wire, one at a time, until the elements all clear those guys. Then, the beam is again pulled up, and the procedure is repeated. **Figure 8-6** shows a big Yagi nearing the top of the tower.

The haul rope is attached to the antenna boom at the boom-to-mast clamp (balance is again extremely helpful), run through a pulley at the top of the tower, and back down to the ground crew. The climber moves upward — with the antenna — while the ground crew carries the weight and lifts the antenna. The moment of some bigger beams will be great, and even though the ground crew will be doing the heavy hauling, the climber will be tasked with some monumental moving of his own to push the antenna around the guys and tower sections.

Antenna design comes into play as well. You must have adequate clearance between the elements near the boom-to-mast clamp, for example, for this procedure to succeed. If the center elements are too close, you may not have enough range of movement to thread the elements around the tower itself, let alone obstacles like guy brackets and guy wires. Temporarily removing some elements may be necessary if you don't have the required clearance on the boom. As with tramming, a well-balanced antenna will make this procedure much, much easier.

I've installed big tribanders, four element 20 meter monobanders and shortened two element 40 meter beams using this method. It's slow, heavy work. In each case, there was simply not enough room to tram them up into place.

Another option that's sometimes used by truly experienced tower workers is to haul the Yagi up as high as possible, secure it, and then attach temporary (steel) guys below it. Then remove the permanent guys above the Yagi, haul the beam up past the guy hardware and secure it. Then replace the permanent guys, remove the temporary guys, and repeat, as needed, moving up the tower. This option is best left to professionals.

Building the Antenna on the Tower

Another option is to build the antenna on the tower and then swing it into position. Depending on the size of the antenna and element locations, the antenna may be built alongside the tower, below the top set of guys, or at the top of the tower. It's a good idea to assemble the antenna on the ground, check balance and become familiar with the hardware pieces and parts before hauling everything up the tower. The ground crew will need to help the climber align the elements. This procedure works well when the boom length is less than the spacing between the guy brackets. **Figures 8-7** and **8-8** show two ways of building an antenna on the tower.

The first step is to haul the boom up onto the tower, with as much hardware on the boom already as is practical — meaning the boom-to-mast clamp and perhaps the element-to-boom plates or clamps. Swing the boom vertical, using either a gin pole or a separate rope and pulley mounted higher than where the beam will eventually go on the tower and rigged near the center of the boom. Then secure the boom to the tower. Once again, those mountain climbing slings work wonders for this; make sure the boom cannot slip out of its vertical orientation.

Mount the largest/longest element first (the reflector) at the bottom of the boom, then work your way to the final director. It's easy to align the elements if you insert a "dummy mast" (a short post or even a piece of PVC, to keep the weight down) into the boom-to-mast clamp, allowing the ground crew to line up elements against it. As the as-sembler, you'll find you're simply too close to accurately judge what's what.

With the elements mounted and aligned, remove the rigging at the director end of the boom and allow the antenna to swing freely. Move the boom so the elements are vertical, and with the ground crew's help, swing them around and into the horizontal plane. Then, the ground crew can lift the beam into its final resting place on the tower as you guide it into place and mount it.

This procedure may be followed if the antenna is side-mounted on the tower, or mounted on a mast above the tower top. The main consideration in this second instance, is that the elements at the head or in front of the boom-to-mast clamp must be able to be moved or swung around the mast. Keep that in mind as you build the beam. You may be able to accomplish this by rotating the elements vertically, then swinging them back horizontal, or you may wish to tilt the boom before installing the forward elements.

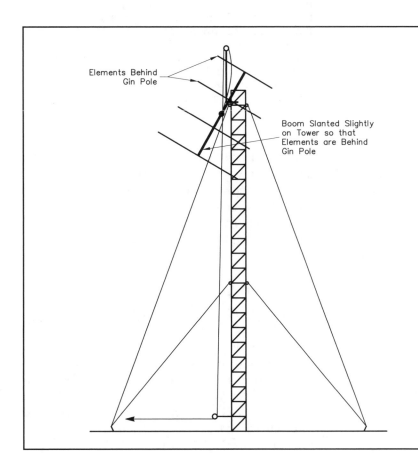

Elements Behind Gin Pole

Boom Slanted Slightly on Tower so that Elements are Behind Gin Pole

Figure 8-7 — Building a Yagi at the top of the tower. The length of the gin pole must be longer than ½ the boom so that the boom can be hoisted upward to the place where it is mounted to the mast. Usually the boom is initially lashed to the tower slanted slightly from vertical so that the top element ends up behind the gin pole. The elements are mounted at the bottom end of the boom first to provide stability. Then the element at the top of the boom is mounted and the boom is moved upwards using the gin-pole hoist rope so that the next-to-top element may be mounted, again behind the gin pole. This process is repeated until all elements are mounted (save possibly the middle element if it can be reached easily from the tower once the beam has been mounted to the mast). Then the boom is tilted to the final position, weaving the elements to clear guy wires if necessary.

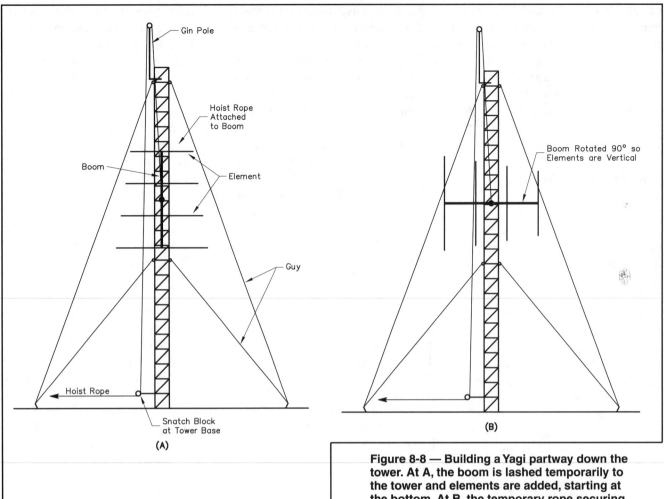

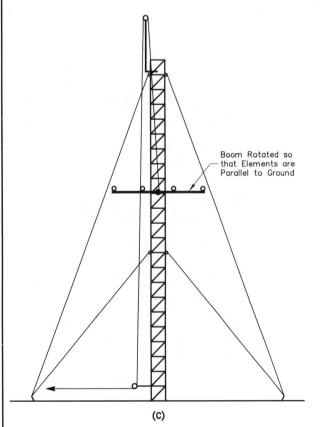

Figure 8-8 — Building a Yagi partway down the tower. At A, the boom is lashed temporarily to the tower and elements are added, starting at the bottom. At B, the temporary rope securing the boom to the tower is removed and the boom is rotated 90 degrees so that the elements are vertical. At C, the boom is rotated another 90 degrees, weaving through guy wires if necessary, until the elements are parallel with the ground, whereupon the boom is secured to the tower.

This narrative all perhaps sounds overly difficult, and it does involve a far amount of climbing up, down, then over and around the elements, but it's often the best answer to getting things on the tower when there simply is not a large amount of open ground space surrounding the tower, which you need for tramming.

The PVRC Mount

The *PVRC Mount* is a tool that provides you a way to build a large antenna at the top of the tower. Named after the Potomac Valley Radio Club (a top contest club centered in the Washington, DC area), it has been presented in the *ARRL Antenna Book* and other publications for many, many years. Its history has been lost among the PVRC brethren, too, although I've tried to research it as best as possible. Credit is certainly due Silent Key members Ed Bissell, W3MSK (W3AU),

Lenny Chertok, W3GRF, and Tom Peruzzi, W4BVV, for contributing to its development. These legendary contest pioneers were all early adopters of massive antenna hardware.

Basically, the PVRC Mount allows you not only to build large beams up in the air, on the tower, piece-by-piece, but also to tilt them over for maintenance purposes. Except for some of the Maryland/Virginia locals, I have never encountered the PVRC Mount in widespread use. Having worked with it many, many times (Frank Donovan, W3LPL, uses it extensively), I can say it truly is a unique and labor-saving device, worthy of any serious antenna builder's attention. See the "Antenna Supports" chapter of the *ARRL Antenna Book* for further construction details.

Coaxial Cables and Connections

Once you have a tower with a mast, beam and rotator installed, it's time to think about connecting it to the radio in the shack, of course. Choosing the right material and installing it properly will help ensure long and trouble-free operation of your antenna system. This chapter is not an exhaustive study of coaxial cable, but rather presents some tips I've picked up over the years.

Figure 9-1 shows the construction of various types of cable described in the following sections.

Coaxial Cable History — What the Letters/ Numbers Mean

Back when the military drove the development of coaxial cable, they designated the cables by descriptions relative to pages in a manual. RG referred to "radio guide." RG-1 was the first cable design, on page one, but the first truly popular cable was RG-6, which is still with us. The designations follow neither logic, nor order, as shown in a few examples in **Table 9-1**.

Since we know that the RG numbers are somewhat arbitrary now, can we assume that say, all RG-58 cables are the same, or might there be differences? If you guessed the latter, you're doing

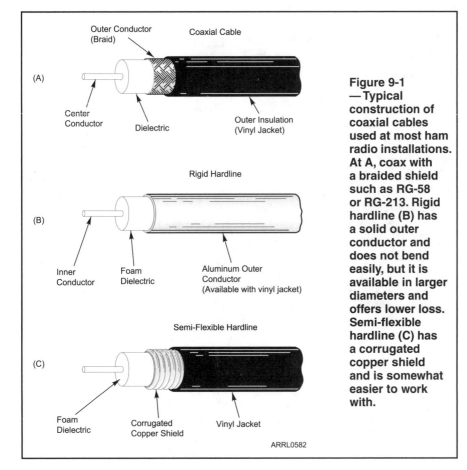

Figure 9-1 — Typical construction of coaxial cables used at most ham radio installations. At A, coax with a braided shield such as RG-58 or RG-213. Rigid hardline (B) has a solid outer conductor and does not bend easily, but it is available in larger diameters and offers lower loss. Semi-flexible hardline (C) has a corrugated copper shield and is somewhat easier to work with.

Table 9-1
Coaxial Cable Designators

RG #	Impedance (Ω)	Center (AWG)	OD (inches)
6	75	18	0.275
8	50	10	0.405
8X	50	19	0.242
11	75	14	0.405
58	53.5	24	0.195

well. For example, is the center conductor solid or stranded? If it's solid, is it bare copper wire, tinned copper wire or silver plated copper wire? If it's stranded, how many strands of what gauge, and from what material are those strands made?

Other issues to question: Is the shield bare copper, single or double braid, or braid and foil? How dense is the braid? How many conductors, and of what size?

Different outer jackets call for different designations, too. A cable might have a jacket made so that chemicals within it will not contaminate the dielectric underneath it — known as a "non-contaminating" jacket.

Finally, after the end of World War II, the whole RG system was simply changed. Coaxial cable today is referred to as C17. To this day, hams have continued to use the old RG numbers, mostly out of habit, although in recent years manufacturers have introduced a wide variety of coax with their own numbering systems.

Why Do We Use 50 Ω Coax?

Lots of folks wonder why or how 50 Ω became popular. The numbers were not arbitrarily chosen.

For pure power handling capabilities, 30 Ω cable would the best choice. We should all know that impedance is the ratio between the center conductor's size and the distance to the shield (braid), along with the dielectric constant of the insulator in between them. It turns out that this swell 30 Ω cable proved exceedingly difficult to manufacture. Another group wanted extremely low signal loss, and it was determined that 77 Ω cable met that requirement. Standard wire sizes and dimensions turned out 75 Ω cable, a suitable compromise.

For those folks who wanted to send very high voltages down the cable, 60 Ω was a good choice. But high voltage users were also likely to be high power users, as well. The compromise between voltage and current? Our ever-popular 50 Ω cable.

Hardline

In today's technology-driven environment, lots of "serious" hams are using large-diameter, low-loss rigid cable — "hardline" — to feed their antennas. It's something to consider if you have to cover a long distance between your station and antennas, or if you want to maximize performance at the higher frequencies. As mentioned in Chapter 6, some hams have found it makes sense to run one very good feed line to their tower and then use a remote antenna switch to select various antennas. You can find data on various transmission lines in the *ARRL Handbook*, the *ARRL Antenna Book* or at vendor Web sites. It's worth taking a look at power handling and loss characteristics and considering options for your installation.

Hardline sizes used by amateurs range from ½ inch (slightly larger than RG-8 size coax) to 1⅝ inch or larger diameter. Construction varies. Some versions have solid aluminum jacket, while others have a corrugated copper jacket (**Figure 9-2**). You'll probably want to look at hardline with foamed polyethylene dielectric. Some versions use "air dielectric" and are designed to be pressurized to keep out moisture. That requires special connectors and equipment and adds a level of complexity not worth the effort for most ham radio applications. Another alternative is coax such as Times LMR600 (0.6 inch diameter) or LMR1200 (1.2 inches diameter), which is built similarly to RG-8 type cable, simply to a larger scale.

If you're used to RG-8 prices, the per-foot cost of these cables and their connectors may come as a shock. Perfectly good surplus and used cable is often available at attractive prices through Internet swap and auction sites. Also, remember that cable television (CATV) hardline is very low loss, and is often available free just for hauling it away. You'll still need to come up with connectors and may need to address its 75 Ω impedance, though. I'll present more on using CATV hardline later in this chapter. Ask around — someone in your club may have a good source for cable.

Working with Hardline

Some of these rigid hardlines have a corrugated outer shield that makes them easier to work with. The two brands you'll most likely encounter are Heliax from Andrew Corporation and Flexwell from Cablewave. (They're the largest manufacturers in the US). Corrugated cables are intended for a gentle bend, but not for continued flexing. In other words, they will bend, but only through a specific radius, and only for a limited number of times. Once that limited number of "flexes" has been reached, the copper jacket will work-harden and then break! The rotation loop will have to be something flexible, like our old faithful friend RG-213. So, run the hardline up the tower leg and stop near the top or at an antenna or switchbox. It sounds pretty simple, but here are some areas to consider before, during and after such installation.

HANDLING IT ON THE GROUND

First, remember the cable will be most prone to damage before you ever get it installed. It's not unheard of for over-zealous forklift drivers to damage the cable unloading it. In case you don't know, hardline comes on spools that are large and heavy. The two ends are often not accessible on new spools, unless you specify that. And the spools are often stored the wrong way — lying flat, instead of being kept upright, and

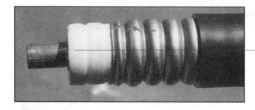

Figure 9-2 — Andrew Heliax is a popular low-loss feed line, but it requires special handling on the tower.

without the factory-installed protective cover, making them susceptible to damage. That's why professionals always use a time domain reflectometer (TDR) to check the cable on the spool before they begin working with it.

Because this cable is so fragile, it's important to consider how you're going to deal with it on the ground. Know where it's going to go and where it's going to come from as it's unwound from the spool. And yes, this cable is delicate and can be kinked, if you're not careful. Have a plan, and work slowly and carefully, following that plan.

HAULING IT UP

Second, the issue of support for the cable as you hoist it must be considered. Naturally, most hams will not have the manufacturer's recommended hoisting grips. (However, this is one thing that can often be found, cheaply, on eBay.) They'll have to improvise. The one thing you *do not* want to do is install the connector, then use that as a pulling point. (Yes, I've seen this done.)

Hardline is prone to damage from jury-rigged clamps and may be permanently deformed or crimped by your attempts to fasten a line. The simplest, easiest and least damaging system I have found (and use) is a choker sling (**Figure 9-3**). A hundred feet or so of this stuff is *heavy*. If you're nervous about the ability of a simple piece of nylon to safely hold this much cable, I suggest you talk to a serious climber — someone who trusts his or her life to such tools. Then (for your peace of mind), secure the sling with some good electrical tape, such as Scotch 88, and haul away. (More on electrical tape in a later section.)

SECURING IT IN PLACE

Third, once you have pulled your hardline up into place and it's hanging there against the tower, the issue of securing it has to be addressed. With RG-8 size cable, you're probably used to wrapping a few turns of electrical tape at regular intervals to hold it to the tower. With hardline, even the best tape often will *not* hold this much weight over time. Like most hams, I cannot always use the manufacturer's hangers. Some solutions I've used include stainless steel hose clamps (both the clamp and worm screw are stainless), spaced liberally throughout the length of the cable, generally every 50 feet. In between, I use black nylon cable ties — those with a 120 pound rating — covered with Scotch 88 tape. I've never observed an interaction problem between the galvanized tower leg and aluminum or copper jacketed cables. (I make a point to ground the cables separately at the bottom of the tower, too.) And yes, you should take this cable's outer dimension into consideration when computing the wind loading you've placed on your tower.

Finally, a few words about maintenance, once you've got the cable installed. Although good quality cable can provide many years of trouble-free service, it does need a little attention from time to time. After years of use, with constant temperature shifting, the dielectric materials can harden, even break. The center conductor can migrate out of center, with obvious bad results. The tighter the bend in your cable, the more likely such problems are going to be. In other words, watch the radius of those bends!

Climb the tower at least once a year and inspect each clamp and tape joint, and re-do any that look weak or sus-

pect. Check the SWR at periodic intervals for each antenna, and compare the results to what you had when you originally installed it. Check all your ground connections. And finally, if you suffer damage, replace the cable. Don't try simply cutting out the "bad" section. Residue from arcing can move down into the cable and create the problem again.

Using CATV Hardline

Hams are always searching for a bargain, whether it's a piece of gear for the shack or a way to save money on the tower. CATV hardline represents one such bargain. This aluminum-jacketed cable is rugged and low loss. For many years, short ends (usually anything less than 300 feet) were often free for the asking from the local cable company. This source is drying up, however, as more and more companies change over to fiber. It's worth checking with your local cable company, or perhaps someone in your club has a stockpile.

Many hams overlook this cable, however, because they know it's 75 Ω coax, and believe using it will cause SWR problems or issues. Even if you use CATV cable with no attempt at impedance matching, the SWR from the mismatch will only be 1.4:1, which is pretty insignificant.

One easy solution to achieving a match to 50 Ω is to use an asynchronous transformer at each end of the 75 Ω hardline. For those with extensive libraries, an explanatory article is "Matching 75-Ohm CATV Hardline to 50-Ohm Systems," by Charles J. Carroll, K1XX, *Ham Radio Magazine*, Sep 1978, pp 31-33. Here's how you solve the problem for monoband antennas (see **Figure 9-4**):

Figure 9-3 — Proper orientation of a sling on hardline — all wraps are flat.

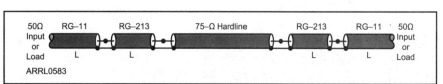

Figure 9-4 — The nonsynchronous matching transformer method of matching 50-Ω loads to 75-Ω hardline as described in "Replacing 50 Ω Coaxial Line with 75 Ω CATV Hardline" by Phil Ferrell, K7PF, Technical Correspondence, Sep 1998 *QST*.

1) Start at the last piece of equipment in your station before the antenna feed line (transceiver or amplifier antenna jack, wattmeter, switchbox or whatever) where you're expecting to see 50 Ω impedance.

2) Install a 0.0815 wavelength section of 75 Ω coax (usually RG-11).

3) Next, install a 0.0815 wavelength section of 50 Ω coax (usually RG-213).

4) Next comes the 75 Ω CATV hardline, any length necessary to get where you want to go.

5) Next, another 0.0815 wavelength of 50 Ω coax (again, usually RG-213).

6) Next, another 0.0815 wavelength of 75 Ω coax (again, usually RG-11).

7) End at your 50 Ω antenna.

Note that these are not physical wavelengths, but *electrical* wavelengths, which must be calculated using the correct velocity factor for your RG-11 and RG-213 cables.

$$L \text{ (inches)} = (961.7 \times VF) / f \text{ (MHz)}$$

where VF is the line velocity factor. Solid polyethylene (PE) is 0.66; foam PE varies from 0.78 to 0.86 for popular RG-8 size cable; Teflon is 0.695.

For example, at 14.15 MHz with PE-dielectric RG-11 and RG-213:

$$L = (961.7 \times 0.66) / 14.15$$
$$= 44.9 \text{ inches}$$

Consult the manufacturer's data sheet for the specific cable chosen, or better yet, measure.

Connecting to CATV Cable

Over the years, a number of methods of making connectors for CATV hardline have been published. Here's how I connect to ¾ inch CATV hardline, which is (or was) very common. See **Figure 9-5**.

The first step to building a connector is to obtain some Amphenol PL-258 double female connectors. (Cheap

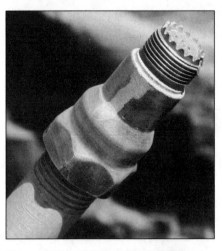

Figure 9-5 — Here's a homebrew connector for ¾ inch CATV hardline. It was described by James R. Yost, N4LI (now a Silent Key) in May 1981 *Ham Radio* magazine.

imports will *not* work.) One end of the PL-258 is held on with a snap-retaining ring, inside the connector. Using a sharp pointed tool, remove this ring. (I use two dental picks. Be careful — these rings can sometimes fly across the room. Patience and practice make perfect.)

Insert the PL-258 metal shell into a copper plumbing fitting used to join ¾-inch OD tubing to ½-inch threaded pipe. Leave a small amount of the PL-258's shoulder (about ½₃₂-inch) above the copper fitting. Now, solder the PL-258 and pipe adapter together. (I use silver solder, but ordinary rosin core will work fine.)

Turn your attention to the parts removed from the PL-258. Using a suitable drill bit or reamer, ream out one of the insulators. Go slow; the plastic is brittle. A lathe is ideal for this, although not necessary if you're careful. Spread the prongs of one end of the PL-258 center conductor so it's a snug fit on the hardline center conductor. Don't do anything to the opposite end.

Insert the smaller end of the reamed-out insulator into the new connector, followed by the spread end of the conductor. Insert the other insulator, small end up, and re-install the snap ring.

On the hardline, with a sharp tubing cutter, cut off ⅝-inch of the aluminum jacket. Do *not* cut the foam insula-

tion. Using a ¾-inch pipe die, thread approximately ¾-inch of threads onto the jacket. Clean the threads, and then trim off the foam (a sharp single-edge razor blade works well). Be careful you don't nick the copper plated aluminum center conductor.

Carefully round off the end of the center conductor. Fill the threads with Noalox or other joint compound and thread your connector onto your hardline.

Install and weatherproof your new low-loss cable.

Looping the Loops

Here are some solutions to the problem of dealing with moving coax — the sometimes problematic "rotator loop."

Probably the most common way to handle the issue of allowing enough room for the coax to turn with the beam is a simple, dangling hunk of cable. Many beams (a typical three element tribander, for example) will have their driven element near enough to the mast to allow you to simply secure the coax to the boom, then leave it hanging loosely straight down so that it travels freely around the mast as you move the beam. The loop hangs out away from the mast or tower, drawing itself up as the beam moves.

The same sort of loop will also work well on long boom Yagis where the driven element is well removed from the mast. In that case, secure the coax along the boom until you reach an appropriate spot close to the tower to form the loop.

Another option, one that works well on flat-top style towers, is to coil the coax in a loop around the mast, such that this loop winds and unwinds as the rotator turns. Of course, this means the outer jacket will actually be rubbing against itself and also rubbing against the tower top as it turns. If the coax rubs against anything (such as the thrust bearing bolts, or the actual edge of the tower top plate), abrasion can occur. Continued abrasion can cause the jacket to tear or wear away, leading to moisture contamination, broken

Figure 9-6 — This swinging gate keeps cables away from rotating tower hardware at the N3HBX contest station.

Figure 9-7 — W3LPL's standoff arm for the rotation loop on his rotating AB-105 tower is simple but effective.

conductors or other problems. I have installed "jackets" over the coax to prevent such damage, using old garden hose or cable loom and the like.

Temperature cycles (winter to summer, for instance) can cause the loop size to change, leading to trouble, too. Obviously, the idea that you have a moving object up at the tower top should cause you to think very carefully about this simple "loop" that you're installing.

Questions abound on what type of coax to use for the loop. I still prefer RG-213 or BuryFlex, as they're tried and true performers. I've had clients tell me they want to use Andrew Super-Flex hardline, as they believe the name implies "super flexibility," but that's a misnomer. SuperFlex really means the coax has a smaller-than-normal bending radius. All helical hardlines have a finite number of bends possible (Andrew says 1000 bends). After too many bends, the corrugated jacket work-hardens and eventually breaks. So a real, braided cable is what you want to install for any loop use.

Side Mounted Antennas

Side-mounted antennas, whether on a swinging gate or an orbital ring rotator, also present special problems. Typically, the simplest solution is to use the truss support on long boom Yagis to hold the coax up and away from the moving parts of the ring. Do not allow any piece of the loop to come in contact with moving parts.

For rotating towers, I have had very good luck using an actual gate hinge holding a "yardarm" out and away at the tower base. The arm is free to turn or rotate with the tower, which helps to relieve any stress put on the coax cables. See **Figure 9-6**. This is especially helpful on rotating towers, which are usually "stacked" full of a variety of antennas, meaning there are numerous cables leaving the tower base. W3LPL uses a simple vertical "post" to hold his rotating loop up and away from his rotating AB-105 towers. See **Figure 9-7**.

Regardless of the kind of installation

you have, it's very important to actually try it — to install the loop, then have someone rotate the antenna through a couple cycles (stop to stop), as you observe what happens. Does the loop move freely, not touching anything? Is there enough coax (especially near the ends of rotation)? Only then do you want to finalize the install — securing the loop permanently. Remember, since these parts will be moving, secure things so that *only* what you want to move can and will move. That means using good, heavy (120 pound rating) UV-rated cable ties or wire ties, for example. I like to cover the ties with Scotch 88 tape for further protection from the elements.

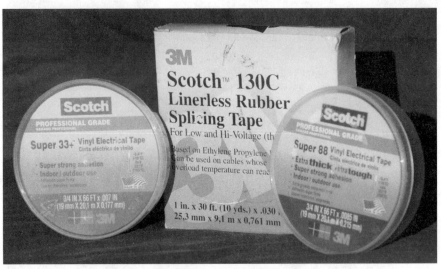

Figure 9-8 — The secret to successful weatherproofing — Scotch 130C linerless rubber splicing tape and Scotch Super 33+ and Super 88 vinyl electrical tape.

Weatherproofing Those Connections

You've studied the options and picked the best coaxial cable for your application, gotten it up the tower and connected it to your antenna. To protect your investment and keep your station working over the long haul, it's a good time to learn about weatherproofing.

Tape Tips

Once, one of my clients, watching from the ground, asked me what I was doing while weatherproofing his cables. "What's that white stuff?" he asked. If you've heard this lecture before, I apologize, but it seems the subject is cyclical. Every few years, it comes back into rotation, showing up on reflectors and the local repeater, in print and conversation.

Weatherproofing (notice I did not say waterproofing) connectors is a foreign concept to some hams. I encounter unprotected connectors all the time, from simple screw terminals to RF connections. There are lots and lots of ways to protect and weatherproof connections that will be exposed to the elements.

Let's begin with tape. Vinyl electrical tape came into widespread use after World War II. Before that, the primary sealing product was a rubbery, sticky substance called "friction tape."

I remember using it as a boy on the farm. I also remember discovering the new, vinyl tape and being impressed. Today, we have a variety of vinyl tapes from which to choose. I believe 3M makes the best, so this article will focus on their products, some of which are shown in **Figure 9-8**. They are readily available at hardware stores, home centers or on the Internet and you should have no trouble finding any of the products discussed here.

Tape Characteristics

Among tape *aficionados*, you'll sometimes hear certain terms or characteristics discussed: adhesion to backing, adhesive transfer, conformability, elastic memory, elongation and flagging. But first, let's consider some real world application or uses for these tapes.

Vinyl tape is used to *jacket* (insulate), *splice* (low voltage connections) and *mechanically protect* (just what it says) joints or connections. Most of the commonly found tapes are rated up to 600 V. The strapping or bundling or holding uses are often taken for granted. For insulation purposes, always use a minimum of two half-lapped layers — meaning each layer of tape overlaps itself by half the tape width.

The final layer is wrapped in a more

"relaxed" manner — without as much tension or stretch being put on the tape. We've all done this when we've wrapped connectors, but chances are, you've not thought much about this characteristic. This ability to fit snugly, making complete contact with the surface of an irregular object without creasing, is called *conformability*. This ability to "stretch to fit" and then "stay in place" is called *elastic memory* and *elongation* by 3M.

All manufacturers grade their tapes according to thickness. Super 33 (7 mil), Super 88 (8.5 mil) and Super 35 (color coding tapes, also 7 mil) are the more popular Scotch products used by hams. Thinner tapes are very easy to apply and will conform readily; they also break or tear quite easily. Thicker tapes provide faster buildup, better dielectric and mechanical strength, and can be harder to cut through.

Scotch 88 can be applied from 0-100°F, extremes under which few of us would want to be working. Scotch 88 also offers excellent elastic memory and elongation, so you can get a good snug fit under a wide variety of conditions.

But is it really and truly waterproof? The answer is no, not always, if used by itself.

The Next Level: Waterproof

There are, however, solutions — additional products that move you from *weatherproof* to *waterproof*. Rubber tapes (not vinyl), which provide electrical insulation and mechanical sealing, can also seal connections against the environment. They readily stretch to conform to irregular surfaces. Quite simply, they form a gasket around cables, connectors or wires. Two readily available products I've used are Scotch 130C or 2242, called *linerless rubber tape*.

Linerless simply means there is no adhesive on this tape. It will *not* stick to connections. Indeed, as you peel it off the roll, it hangs limp in your fingers, and you're not sure it's going to be any use whatsoever. It does, however, stick to itself (called *self-amalgamating*). It should be wound tightly around connectors — starting from the bottom up like roof shingles — keeping it under tension, as you go. Scotch recommends a 50% stretch, and with some practice, you'll get the hang of how much stretch this means. Experience will be your guide, but one good rule to follow is when the tape turns from black to gray, you should stop stretching. It stretches and molds itself easily around connectors, like PL-259s.

Building a Waterproof Connection

Figure 9-9 shows the steps to making a long-lasting waterproof connection. In practice, I like to wrap connectors with some special Teflon tape first. It's 0.003 inch thick. At first glance it looks like the ordinary pipe thread tape plumbers use, but it's tougher stuff and conforms to literally anything! It's available from McMaster-Carr. Then comes a layer of Scotch 130C linerless rubber tape, followed by two layers of Scotch 88. An alternative is to use

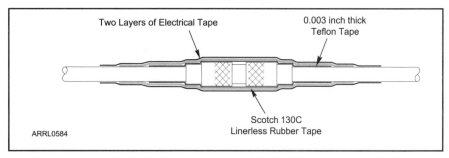

Figure 9-9 — A Teflon tape base layer, followed by a layer of linerless rubber tape and then two layers of vinyl electrical tape makes a long-lasting waterproof connection. See text for details.

Scotch 88 rather than Teflon tape for the base layer.

As to technique, it's old news, but the key is to make sure there are no gaps or air cavities in your wraps. Where there is air, there will be water, plain and simple. The super-flexible Teflon tape will literally conform to anything, solving that problem.

On that last wrap of Scotch 88 tape, let the end relax, and then *cut it*. Don't simply grab the tape, tug, stretch and tear it. Over time, the elastic memory will cause the tape to become loose and *flag* (that's what those nasty little flapping ends are called).

Think about each joint being laid like the shingles on your roof — each overlapped layer is over the next layer, as you move the joint. Where practical, I also like to put a tie-wrap above each connection, where capillary action will

Figure 9-10 — Coax connectors removed from service at N4ZC after 12 years of weatherproof and waterproof service.

pull the water away before it gets to run down into the taped joint.

How well does this technique work? It will take a few moments of effort to get a knife or blade sawing through that connector after a couple of years, but once you get an end loose, you can peel the whole mess away and be surprised with a brand-new-looking connector. It's somewhat expensive (I buy mine at Home Depot), but the ability to truly waterproof a connector over an extended period of time is invaluable. **Figure 9-10** shows what a coax connection protected this way looks like after many years of service.

Scotch Liquid Electrical Tape

Liquid electrical tape is another quality Scotch product, although I've only recently begun using it. If you have the time to wait for it to set up, and the ability to brush it on to the connector (for instance, if you're not hanging upside down and backward 120 feet in the air), it works perfectly to seal the 130C-tape-covered connector. It's also a good choice for covering normally exposed antenna electrical connections — such as small feed point hardware. I've been impressed in the short time I've been using it. Try this liquid electrical tape for such joints.

Inspections and Maintenance

Once your tower is standing tall and the antennas are up, go ahead and enjoy the fruits of your labor. Keep in mind, though, that your job is not finished. Regular inspections and maintenance are the best way to keep your station operating at top performance and head off small problems that could turn into big headaches. Please understand that this is an ongoing commitment, and this chapter offers some tips for keeping your antenna system working efficiently.

I suggest to my clients that they perform an inspection (or have such work done) twice annually. Springtime is one obvious choice of an opportune time —

the tower(s) and antenna(s) have just gone through the temperature cycles of winter, along with stress from possible ice loads, high winds and the like. Fall is a second obvious choice — right before the start of contest season. Regardless of when you do it, inspecting your installation should be appear somewhere on your own personal "to do" list.

A Checklist Approach

While I've often joked, saying that this is "one job where you can start at the top," that is not literally true. Any tower inspection has to begin on the ground, of course. And you begin by examin-

ing the tower base and guy anchors (if applicable). You are looking for any hardware that might have come loose since the last inspection and fallen to the ground. You are looking at the concrete and base area for signs of winter weather damage (cracks and the like).

Figure 10-3 — Tower legs can rust from the inside out. Note the corrosion on the leg and bracing rod where the steel entered the concrete base.

Figure 10-1 — After 30 years of service, this guy anchor had severe corrosion from contact with the soil. The anchor looked fine above the ground, but a little digging and close inspection revealed a failure waiting to happen.

Figure 10-2 — This close-up shows how the lower end of the anchor had practically dissolved in the ground. The rusty steel is crumbling, with very little strength left to anchor the tower.

Figure 10-4 — On the outside, this tower leg appears to be fine. The galvanizing is intact and showing only minor staining. Inside the leg it's a different story.

You are looking at the anchors where they leave the ground. It's not a bad idea to actually dig up the soil surrounding those anchors just a bit, to see if there's evidence of corrosion or lost steel below grade. **Figures 10-1** through **10-4** show some of the problems that may be lurking below the surface.

With the ground inspection complete, I'm climbing to the top of the tower and working my way down.

- All RF and control cables fastened to the tower will be inspected.
- Electrical connections to rotators and antennas will be inspected.
- Tower bolts and nuts will be visually inspected for tightness.
- If needed, I'll retape or secure lines, along with coax and control cables as I go.

While this may sound overly complicated, once you've done it, it becomes almost second nature to climb up, take a quick look around, tug and pull on various things, then begin working your way down. Even a tall tower can be completed relatively quickly, once you're familiar with things. And if it's your own stuff, you will be, or should be, familiar with it!

Keeping Track

All this searching and looking should be recorded, preferably in the Station Notebook (you *do* have a station notebook, don't you?). If there is evidence of possible physical damage, or if something shows early signs of deterioration, it's a good idea to document it with some pictures so a visual record exists for later comparison.

The sidebar details my thoughts for a Station Notebook. The section dedicated to Towers has entries for various measurements and data, along with some procedures and guidelines. **Table 10-1** shows some topics to include.

The Station Notebook should be fleshed out with details for your installation. For example, some of the activities or procedures might read:

1) Each tower base will be checked for rust or other deterioration of steel and any accumulated dirt or other debris will be totally removed.
2) Tower plumb to be checked from top to bottom (using a transit, or a pair of them), and corrected by re-tensioning the guys as needed.
3) All tower ground wires and ground rods to be inspected.

While the simple list presented here is not as complete as what you should have in your Station Notebook, you can see what we're getting at with this sort of recorded detail. What we're looking for is a systematic examination of various mechanical and electrical items, with places to describe what's happened over time and note what needs attention. Writing it down makes sense. You *will* forget details as time passes, so having a written record makes your job much, much easier.

And while the example shown here is my old, original personal Notebook, collected in a three-ring binder, there's something to be said for keeping such files on a computer instead. I suggest this mostly due to the ease and convenience of digital photographs these days. Documenting things has never been easier, and since I've acquired a laptop, ostensibly to use in writing this book, I've taken to storing and recording a wide range of things on it. I imagine I'll have the Station Notebook there by this time next year!

Some Thoughts on Tower Steel Maintenance

Most towers hams own are made from steel. Most are protected from environmentally induced corrosion in one of three ways: paint, hot dip galvanizing, zinc coating or some combination of these. Paint is considered the least effective. It depends upon the paint forming a protective coating to seal the steel from moisture and other corrosive agents. If chipped, the steel underneath will begin to corrode. Hot dip galvanizing is the most widely used protective method. You've heard of it — it was part of Rohn's magazine advertising for years.

Hot dip galvanizing means thoroughly cleaning the steel, then submersing it in a molten zinc bath. When removed, the zinc forms a thick, abrasion-resistant coating that is metallurgically bonded to the steel. The zinc coating protects three ways. First, the coating is an effective barrier against water and corrosive agents. Second, the zinc dissolves to provide cathodic protection to any exposed steel. Third, over time, the zinc forms zinc carbonate, which seals over any damaged areas in the coating. It's this cathodic protection that differentiates this coating method from all others. Zinc is more anodic than the steel it covers. Small scratches or cuts in the coating mean corrosion products precipitate

---

Table 10-1

Tower and Antenna Inspection Checklist

Tower Base
Concrete: Yards used in construction
Maintenance performed

Anchors
Concrete: Yards used in construction
Maintenance performed

Guy Wire Tensions
Set 1
Set 2
Set 3

Turnbuckles
Safety wires
Preform Guy Grips or Crosby Clips

Ground System Condition & Lightning Protection System
Ground rods & wires

Tower Hardware
Check tower for plumb
Guy brackets

Antennas
Installation date(s)
SWR recordings/band

Cables & Electronics
Installation dates
Length, color codes, connectors

The Station Notebook

With apologies to Henry David Thoreau

I'm often surprised to discover that fellow hams (or clients) do not have, nor keep, notebooks concerning their stations. Something like a naturalist's set of observations, if you will. Now, a hurdle for today's active ham can be the high cost of gear, including test or recording equipment. Yet, one of the most useful tools available costs next to nothing — the simple station notebook.

Why bother? A number of reasons, of course, but here are a few:

1) You can and will forget things.

2) You will have data to compare when things go wrong.

3) You will learn more about your station because note-taking will actually improve your ability to observe.

I've often heard folks say they like to use the backs of their station logbook pages for this sort of thing, and I used to do that myself. But it quickly became apparent this sort of record keeping was simply too random — there was no easy way to find or retrieve any information once I'd recorded it. Obviously, what I needed was something a bit more organized and serious.

And this will probably come as a shock in today's digital-everything world, but I still like a hard copy instead of electronic formats. I find it's faster, more convenient, less prone to loss or theft, and way more portable than any computer-oriented system. Use index cards, loose-leaf notebooks, a sketchpad, or whatever works for you. Check out stores that sell artist, school or office supplies for suitable solutions.

I like the three-ring loose-leaf notebook approach (**Figure 10-A**). I use regular lined paper for notes, along with graph paper for drawings or sketches. The standard 8½ × 11 size means I can insert copies of articles (where relevant), too. Use a waterproof ink pen, not pencil, for notes and such.

What are some things to include, or how can such a notebook be truly useful? In no particular order, consider several tabs for various topics or subjects. I have sections for:

Serial numbers. When I purchase gear, I record relevant data here, including model and serial numbers. Everything is in one place, which makes preparing a list prior to overseas travel simple and fast. And it's also a great insurance tool (although one I hope you never have to use, and I'm speaking from experience here!). I include information from the supplied manual as well, in case it's later lost or misplaced.

Antenna system. This is the largest file, and perhaps the most important. Besides the usual SWR or impedance readings and descriptions, I include data on the type of coax or feed line, including date installed, information on connectors and the color code used. As changes occur, these get recorded as well.

Towers. This is the second largest file, with installation data and maintenance data. It includes recordings of guy tensions, hardware used, relevant data on tower bases and guy anchors, and so forth. Once or twice a year, when the

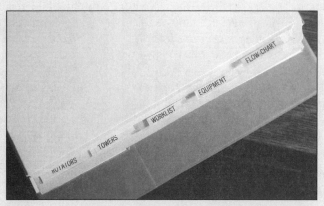

Figure 10-A — My Station Notebook includes detailed information about my towers and antennas. Having a record of installation dates, materials used, maintenance performed, lengths and dimensions, electrical and RF cables and connectors, "normal" measurements and so forth can be invaluable when tracking down a problem.

tower inspection rolls around, it's especially handy to be able to see what changes have occurred or what repairs need to be done. Again, as changes take place, they get recorded here.

Rotator system. Similar to the antenna file, with voltage and resistance readings, color codes, and other pertinent data.

Station signal flow-chart. A block diagram of every signal (RF, AF, logic and keying or control) or circuit path in your setup, this is the type of thing you often see in *NCJ* or *QST* articles. It's amazingly helpful when things go wrong or stop working to be able to know exactly (especially when you're suddenly frustrated) what each box is supposed to do. All the cable IDs or label tags get recorded here, too.

Work list. The one area where I'll allow myself to throw things out afterward, or otherwise be sloppy. It's one ham's approach to those silly "things I gotta do" notebooks you see for sale in discount stores. In one place, I can map out or write down project ideas, vendor information, contact information, reactions from fellow contesters or ideas gleaned from articles or the Internet. In short, it's a catch-all area, a place where I plan for the future. The key factor for this tab is that, once it's written down, I have a much greater tendency to recall or remember it!

Such recordkeeping and summaries of your station building, along with operational notes, will increase your ability to troubleshoot and repair the system when it breaks. Trust me — this is a good idea. And, I'm willing to wager, you'll feel some further manifestation of pride or accomplishment in making such journal entries or summaries. A little glance backward, toward Thoreau, who said, "I feel as if my life had grown more outward when I can express it." It's a perfect metaphor for radio's ability to express and explore, as well.

---

on the steel surface. Corrosion will not occur between the zinc and the base steel. (If you remember high school chemistry, you know we're talking about a galvanic cell.) In hot dipping, several layers of zinc form — 100% zinc at the surface, ranging through various layers of alloys to 100% steel at the base.

Hams often attempt painting maintenance on hot dip galvanized towers, with varying degrees of success. Here's why. Upon removal from the dip, the zinc coating begins a variety of chemical transformations. The outer layer (that 100% zinc) oxidizes. Zinc oxide, zinc hydroxide and other surface conditions can be present from

Figure 10-5 — Zinc-rich cold galvanizing compound (spray paint) for touching up hot dip galvanized tower parts is generally available from hardware stores and home centers. Be sure to get a product specifically made for galvanized surfaces, and not regular paint.

Figure 10-6 — The galvanizing around the bolt holes in this Rohn tower leg had become worn over the years. The underlying steel was still solid, so a coat of zinc-rich paint helps to minimize corrosion and extent the life of the tower.

48 hours up to several years after galvanizing. In any case, this oxide prevents paint from adhering to your tower. Light sandblasting or a high-pressure wash followed by a metal conditioner is required for proper preparation — usually not too practical, regardless of whether you've got a 200 foot or 50 foot beauty in your backyard. A simple cleaning with vinegar will provide the simplest and cheapest approach to helping your paint adhere to a hot dip galvanized surface, though. Try that, before giving up.

ASTM standard A123 covers hot dip galvanizing. Basically, your inland-USA tower should last at least 30 years with a typical 3.4 to 3.9 mil thick zinc coating. At a minimum, your tower will require touch-up after erection. Over time, you may notice rust stains or corrosion. What you're probably seeing is rust from leg bolts or other hardware you've fastened to the tower. Or something may have damaged the galvanized coating in some fashion.

What should you do? Touch up those parts or areas of your tower with zinc-rich paint such as the Rustoleum, ZRC or Brite Zinc products found in many hardware stores and home centers (**Figure 10-5**). Read the label to guarantee you're getting good paint and expect to pay more for the proper paint. Read the label again regarding preparation and application. Most of these paints are not intended to be applied or used in periods of high humidity — usually just when we're thinking about doing this sort of maintenance. However, there's really no alternative.

Most experts will tell you that quality paint, properly applied, will adhere well and will furnish protection, assuming some zinc remains to provide cathodic coverage to the steel base of the tower. So, with some surface cleaning, a good zinc-rich paint is indeed a solid maintenance method. **Figure 10-6** shows touch-up of bolt holes at the bottom of a tower section.

Putting it all Together

The previous chapters have covered the various considerations, techniques and tools for planning and installing a ham radio tower. The narrative, or story, of this chapter suggests how it might all come together for your first tower.

A Tower for the Taking

One night, after supper, your phone rings, and the caller is Ed, your ham buddy across town. He knows you've been wanting a tower, and he has a line on a tower that's available free, provided you take it down. It's located at an old taxi company, and the owner of the building would like it removed within the month.

Ed's not too sure how high it is, but he thinks it's 50 or 60 feet tall. He promises to help you take it down, and he is sure some other club members would be willing to help with this work, as well.

Your question: Is it worth pursuing this, or not?

A couple of nights later, you stop at the address on your way home. The tower appears to be Rohn 25 with two small vertical antennas, one mounted on the very top of the tapered top section, and another about 10 feet down. Heliax feed line runs up the tower

to both antennas; you're pretty sure you could use that, too. You count the sections. It's a 50-foot tower, all right. There's a house bracket near the eaves of the building, about 25 feet high. And there are three guy wires, at what you've read is the appropriate 120-degree spacing. They look like the right stuff, EHS (Extra High Strength cable). Two go to elevated guy posts, made from I-beam material. One goes to a utility pole and is attached to a very large eyebolt, secured through the pole, apparently to allow clearance over the driveway.

The sections all look straight. No rust is visible; indeed, the galvanizing looks good, considering this tower has been here for at least 10 to 15 years. Ed said the building's owner told him the tower had been put up by a company here in town that has since gone out of business, but they were real professionals, having had something to do

> *"Never having had a tower before, you realize you don't have the necessary safety equipment, nor the knowledge required."*

with the local AM broadcast station and their tower.

What Will You Need?

You have to decide, and soon, whether you should pursue this, or not. Having read a bit about towers, you know there are certain somewhat specialized tools you will need to take this tower down. A gin pole, first of all, allows you to stack or remove the 10-foot sections one at a time.

To get the sections apart, everyone recommends something called a TowerJack. Then there's the rope required — at least 150 feet of it, to work up and down the tower, in and out of the gin pole, with a comfortable working length on the ground. You'll probably need some pulleys, some hand tools and a small trailer to haul the sections home (you know you can borrow Ed's). Then there's the question of time, gas used in ferrying things here and there, and so forth.

Never having had a tower before, you realize you don't have the necessary safety equipment, nor the knowledge required. You're probably not ready to climb this one. At least not within the span of one month — the time required in which you must remove the tower. So, you decide to get someone else to help you, to do the work you are not yet prepared to do.

You contact Gary, one of the "big gun" DXers in your local radio club, who has three towers at his home. Even though Gary's retired, a distinguished-looking white haired gentleman, you know he still climbs his towers. You ask him for help.

Gary spends nearly an hour with you on the telephone, going over the entire process of taking down such a tower a couple of times, explaining not only how the work should proceed, but explaining the *why* behind several of the procedures. He agrees to help you on Saturday morning, providing the weather cooperates.

Work Day Arrives

Saturday dawns with temps in the 50s, the sky already blue and bright.

Gary calls while you're grabbing a bite to eat, to assure you he'll be heading to the tower site momentarily. You realize it's even further away from his QTH than yours, which makes you feel guilty. You make a note to stop by a donut shop for coffee and goodies on your way to the site.

When you arrive, Gary, Ed and a couple other helpers are already there. Gary's a retired manager at a local electrical utility, and his calm command presence puts you at ease right away. He's already explained to Ed and the others what they'll be doing today; it's just what he described to you the other night on the telephone.

Gary digs at the base with a small shovel, telling you he wants to confirm the base is not only the proper size and depth, but also still solid. Then, he walks over and inspects each anchor post, paying particular attention to the utility pole. He tugs on an elaborate-looking bright blue-colored web contraption, festooned with hooks and rings, then steps up and onto the tower. He reaches up, snaps a large hook over his head and climbs up, repeating the process as he goes. You notice he has another loop around the tower at his waist. You also notice he's carrying a thin rope attached to that belt. He stops

and spends considerable time at the house bracket, obviously inspecting it closely, too.

Then, he climbs to the guy bracket, and stops there. Satisfied, he drops the thin rope to the ground. Ed takes the already attached clip and hooks on to the gin pole lying on the ground, then passes the other side of the rope into a pulley that's clipped to the tower. *Wait a minute...how was that possible — without having the end of the rope?* Ed smiles, and shows you the pulley, which rotates apart, allowing you to attach it at any point on the rope. "Gary has some really cool tools," he says. Then he steps back away from the tower, pulling the pole up, slowly and steadily. When Gary hollers, Ed stops, and soon the gin pole is attached near the top of the tower.

Next up the tower is a small bag of tools, and before you know it, Ed has you carrying a small stepladder to one of the elevated guy posts. Gary has left a spray can of lubricant, with instructions to use it on the turnbuckles and bolts. You do, and then slowly work the turnbuckle back and forth, until it's turning freely, and loose, until Gary says stop. You repeat this process twice more at the other anchors.

By then, you see Ed's at the base of the tower, helping Gary lower one of the vertical antennas. The second one follows in short order, then the Heliax feed line. You didn't notice it, but apparently Gary loosened or cut the tape securing it to the tower legs as he climbed up. Everyone works carefully to make certain the fragile feed line isn't damaged or kinked, coiling it into large-diameter reels once it's on the ground.

You see Ed and the other two helpers working at taking one guy wire loose totally. You walk back near the tower base, relatively frightened when you realize that Gary's hanging on up there with no guy wires attached. But you notice that the tower's not moving very much, and you figure he must know what he's doing. As if he's reading your mind, Gary says, "Since we've only got two un-guyed sections above the house bracket, we can safely take

down the top section after taking off these guys."

And seemingly before you know it, that's exactly what's happening. With very little apparent effort on anyone's part, the topmost section is up and separated from the other sections, dangling there on the gin pole rope. Three of you carefully lower it to the ground. You're surprised at how easy this is, but then, Gary's probably done this dozens, perhaps hundreds of times, and he has the tools. Besides, the section only weighs 35 pounds... but you're excited. It is, after all, your first tower!

The takedown proceeds quite smoothly, and soon you're staring at five pieces of tower loaded on Ed's trailer. He even offers to haul it to your house. Gary asks if you've figured out where it's going in your backyard, and you own up that, no, you've looked around the yard several times, but reached no final decision. He offers to come over and look at things, and you offer him lunch. He accepts and you reach for your cell phone, checking with your wife....

Planning Your New Tower

On Monday evening, after work, Gary drives over to your house. You chat about the yard itself, your shack location, and what your family has told you. He has a fiberglass tape measure, and a simple layout tool, a piece of plywood that has a tower and its associated 120-degree guy spacing drawn on it. Using it, he paces out the orientation possible (behind the garage) and then measures the distances for the guy anchors. He puts you at ease, talking about his own first tower and beam years ago, sharing some stories and asking what you want to do with your own first tower. While you feel a bit awkward talking about DXing with someone who's worked everything, everywhere, several times over it seems, Gary is pleased with your enthusiasm and assures you there'll be great times ahead, once the beam goes up. He outlines what you should do next, the things to have with you and so forth, when applying for the usage permit (all your county requires) for the tower.

So, with all that in mind, you arrange with your manager to possibly come back a bit late from lunch on Wednesday, and drive to your county zoning office. It takes a few minutes to find the right department, but you go in and ask what you need to do to obtain a permit for a ham radio tower. In your briefcase are several items Gary suggested you bring along — the relevant pages copied from his Rohn technical catalog, showing the base construction for your tower, the guying requirements, and the plat drawing of your property, along with another drawing showing where and how the tower will be placed on the property. The guy behind the counter yawns, tells you that you don't need a building permit. But you persist, and explain exactly what you *do* need or want, and before long, the use permit is issued to you.

You're on your way!

Working in the Dirt

Having secured the necessary permit, you know the next step will be digging a hole for the tower base and guy anchors, and then pouring concrete. Gary has mentioned to you that the original installation used a base section designed specifically to be buried in concrete, which you've left in the ground at the old taxi service company location. Therefore, you'll need something to replace that and also you'll need to buy three guy anchor rods.

You are surprised to find a section of Rohn tower lying next to your garage one evening after work. E-mail from Gary explains that you'll simply bury part of this section in the ground as your new base, replacing the originally-buried Rohn base section. The man's generosity and willingness to share amazes you once again.

His e-mail contained the phone number of a backhoe guy, so you call, and arrange to meet him one evening after work. Turns out he's not a stranger to digging tower base and guy anchor holes, having worked for Gary previously. He understands the importance of creating a smaller hole, and one that's square. He mentions he'll be using a smaller bucket to dig your base. You arrange that for the coming weekend.

Gary doesn't leave you hanging on this project, coming over once the hole is dug to guide you through the construction of a simple timber 2×6 framed form surrounding the hole. With his help, you visit the local home center and purchase some rebar, along with some other "essentials," from a list he's carrying. Later that day, he shows you how to make a simple cage for the base and anchor holes, and before you know it, you're ready for concrete.

Turns out, one of those extra helpers from the takedown project is a mason. He gets you a better deal on the concrete — using a Ready-Mix factory truck to deliver some 4000 PSI mix. You've done some Internet research and you think you can smooth out the concrete. Gary has set and plumbed the tower section in the hole, including several inches of gravel at the bottom of the base hole, for drainage of the tubular legs. Then, he'd secured it (wired to some 2 × 4s) so it cannot move out of plumb. The Ready-Mix driver is cautious, and eases the heavy concrete into the hole. Before long, you have the base smooth and looking fairly professional (even if you do say so yourself). Next are the three earth anchors, which are smaller and easier. You pay the driver, and he asks you what you're building. "It's a radio tower. My first," you explain.

That night, Gary calls on the phone. You talk about the process of allowing concrete to cure, and how to proceed with the remainder of the project.

Stacking Steel

One night, Gary calls and says it's time to actually begin putting the tower sections together. Again, he's already organized things, if your schedule will permit, and so forth.

You agree, and the weekend brings help from the same fellows who helped with the takedown.

Gary has a small section of plywood on the ground near the base section, and he's examining each leg of each section — peering down, saying it's important there are no obstructions (wasp or spider nests, clumps of dirt, and so forth) lodged there. Anything that could allow water to collect and possibly freeze and crack the tower leg must be removed before assembly.

Many of the same tools used at the takedown are in evidence, even the TowerJack. You thought that was used to separate sections, but in a few deft moves, Gary shows you how it also works in reverse — pulling the Rohn sections together using the Z-bracing. Pretty clever. Soon, he has put together the four sections, and marked them with some colored tape bands, assuring that what mates well on the ground will go easily together in the air. Again, simple but pretty clever.

Soon, he's once again stepping up and on to the first section, cemented in place, attaching his gin pole, and almost before you know it, there's your very first section of tower, bolted into place, and looking mighty good. Gary mentions that the piece of plywood helps keep each section out of the dirt as it's prepped to be hauled up on the gin pole. He shows you how he puts a small amount of grease on the swaged siderail legs, and how he does not tighten the bolts so much that the tubing is crushed. This will ensure the tower can easily be taken apart later, if need be. You realize, once again, you're learning a lot.

The ground crew guys are used to this process, obviously, and they have an easy-going rhythm to the work, laying out the shiny new EHS guy cable, then pulling it up to Gary on the tower. You watch as they install the Preform Big Grip Dead-ends, marveling at the ease and speed with which these devices make short work of getting the guys on the tower and attached to the guy anchors. The bronze-looking clamps used to hold the guy wires in place temporarily, you learn, are called Klein Grips, and they're another clever tool.

Shortly after lunchtime, your tower is finished, at least in terms of having all the sections assembled, the guys attached, and everything tensioned and secured. Gary suggests that waiting a week would be a good idea — giving the guys a chance to stretch, the hardware to cycle through some temperature changes, and so forth. Staring numbly up at what seems a very tall tower, you mumble something that sounds like thank you. Then, you realize that maybe it would be a good idea not only to thank everyone, but to provide lunch (as intended), so you hurriedly suggest some burgers and franks from the grill will be forthcoming in short order.

Some Final Thoughts

Having put up your very first tower, with the help of local radio club friends, especially Gary's, which included even buying and installing one of his old triband Yagis, you've learned a tremendous amount. You've been on the air more often (despite some

> *Indeed, the whole process of knowing when certain bands are open and why has become totally fascinating to you.*

disappointing propagation), including experiencing some success on the low bands from the wires mounted on the tower. You've even experienced your first contest, something you'd never tried before (and had fun!).

One night, after dinner and the local news, you found yourself sitting in your armchair, looking out toward the tower and antennas, contemplating some of what the experience has taught you.

You've learned something about mechanics. You've learned something about construction. You've learned something about the physics of radio signals. Indeed, the whole process of knowing when certain bands are open

and why has become totally fascinating to you (Gary suggested all along this would happen.). You've learned something about several of your local radio club friends, too. Without their help, the entire project would never have happened, after all. You've learned some things about your neighbors, your local government (not all of that was pleasant, but learning is sometimes painful). You've learned (you think), even more about what it means to be a ham, and feel more connected to the history behind that, in some small way.

The very process of putting up the tower, learning how to climb it and work on it, and using the antennas you put up — all of that has made you, if not a better ham or person, at least more aware of a great many more things than you first thought possible.

That's pretty amazing, you realize. Pretty amazing, and special, too. You wonder what's on the bands right now, and head down to your basement ham shack....

12

Insurance

Having invested some time (just in getting to Chapter 12, as well as in the building of your station!) and perhaps some considerable funds in your antenna system, it's only logical to consider insurance. As a homeowner (or even as a renter), you do have some choices when it comes to insurance. The first place to start is with your own personal insurance agent.

I've interviewed some agents (locally, online and at the Dayton Hamvention), and their overwhelming reaction was disbelief that the homeowner did not read, understand or appreciate their own coverage.

In every instance, those agents had horror stories about owners who thought they were covered, but were not. In every instance, this lack of knowledge among policyholders not only disappointed them, but also saddened them. (All of the agents mentioned they want to finalize any and all claims quickly and smoothly — they want to get them over with!) So, talk with your local agent, and follow his or her advice. Don't guess, or simply hope that you're covered.

After all this research, it became apparent I needed some professional assistance to adequately prepare this chapter. I was determined to make this chapter not only technically accurate, but also useful in some advisory or preparatory way — my goal for the rest of the book.

With that in mind, I contacted Ray Fallen, ND8L, a State Farm Insurance agent in Hubbard, Ohio. Ray wrote an article on insurance that appeared in the February 2009 issue of *QST*. After a series of emails, and an extended discussion at Dayton, I asked Ray to expand on some of the topics from that article for this book. I'm pleased that he obliged.

Homeowner's Insurance:

Your Antenna System and Your Gear

By Ray Fallen, ND8L

Many years ago, I was just like you. I'd buy an insurance policy to protect my home, my car, my life or my health. While the agent tried to explain it, my eyes would glaze over and my breathing would slow. I would nod my head knowingly, make all of the appropriate "Why Yes, Mr. Insurance Agent, your Amazing Explanation just couldn't be clearer" noises…and in my head a voice was whispering the Novena of the Newly Insured: "If something happens…I sure hope it's covered…for I have not one single clue what this guy is trying to tell me."

Then, on February 1, 1988, I went over to the Dark Side.

I became an insurance agent.

This chapter in Don's book shares some of what I've learned in that time about homeowner's insurance, your antenna system and other ham gear.

Caveats and Stuff

Like everything from bulldozers to your morning oatmeal, this chapter comes with some disclaimers:

1) Most homeowner's insurance policies written in the US (except in Texas, and nobody knows why) are based on standard language provided by the Insurance Services Office. Each insurance company modifies that language to comply with individual state insurance statutes … so coverage that applies to towers

in Ohio may not necessarily apply to towers in Florida.

2) Coverage varies from company to company, state to state and country to country.

3) Because of these variations, this chapter can't tell you *exactly* how your antenna installation will be covered. I can give you some talking points to follow when you visit with *your* insurance agent.

And you *are* going to do that real soon, aren't you?

Let's begin at the beginning. An insurance policy is a legal contract between you (The Insur-ed) and the company (The Insur-er). The contract is quite specific in its definitions, coverage, and the duties and responsibilities of both parties. For example:

1) What's covered in the policy is in the policy contract.

2) What's *not* covered in the policy is in the policy contract.

3) Your duties following a loss are in the policy contract.

4) How the loss will be paid is, you guessed it, in the policy contract.

If you're starting to see a pattern here, go to the head of the class.

Here's the problem: I said the policy was specific…I didn't say it was always easily understood. Insurance policies are legal documents written in an obscure Olde English dialect called Lawyer. You need to visit your agent, who can translate Lawyer into English. If you're planning an installation, see your agent before you start. If your tower's already up…what are you reading this for? Go see your insurance agent…*right now*!

Property Coverage

Damage to towers, guy wires, rotators, tower-mounted antennas and related cabling is covered by Dwelling Extensions, Other Structures or Appurtenant Structures coverage listed on the Declaration Pages of your Homeowner's Policy. (The specific name and amount of the coverage varies by company.) For example, State Farm's

Ohio Insurance Homeowner's Policy defines Dwelling Extensions as "other structures on the residence premises, separated from the dwelling by clear space." Hmmm…starting to sound like Lawyer, again.

Dwelling Extensions are *permanently* attached to or otherwise form a part of your property, but they are not *physically attached* to your home (dwelling). Dwelling Extensions are typically covered for 10% of the dwelling coverage amount. For we mere mortals, this may be more than adequate. At large contest stations with multi-kilobuck tower and antenna systems such as K3LR, K8AZ, W3LPL or NR5M, that may be an entirely different story.

If your tower is physically attached to your home (dwelling), then the cost of the tower and related equipment should be added to dwelling coverage on your policy, *not* dwelling *extension* coverage. If there are questions in your mind, this would be an excellent time to have your insurance agent drop by the house and take a look.

Remember, tower replacement costs should include *professional* help in removal, repair or replacement, in addition to the replacement cost of the damaged gear. Also, consider *other*

dwelling extension items: fences, sheds, pole barns, gazebos, detached garages, flagpoles and in-ground swimming pools, for example. Add up the *replacement* cost of those structures and if you need more coverage,

Hurricane Ike sent this Comet GP3 fiberglass vertical right through the eave. (W5JON photo)

In the wake of Hurricane Ike, W5JON's antennas looked like this. Can you find the SteppIR in there? (W5JON photo)

buy it…now. If you're not sure of the *replacement* cost of those structures, your agent can help.

Wire antennas in trees and/or ground-mounted verticals are generally not permanently affixed to your property (sometimes much less permanently than we'd like) and are considered personal property. The good news is personal property is covered for 50% to 75% of the dwelling amount…the bad news is the coverage is not as broad as dwelling extensions, but wind, ice, vandalism or lightning damage is usually covered. (Review the discussion of named perils in the ARRL Ham Radio Equipment Insurance section which follows.)

The distinction between dwelling (and dwelling extensions) and personal property is:

- If something's "nailed down" and not easily moved, it's dwelling or dwelling extension coverage.
- If it's easily moved (even if it's "nailed down"), it's personal property.

Believe it or not, most insurance companies look at in-ground swimming pools as dwelling coverage, but above ground pools are considered personal property, since they can be

Hurricane force winds toppled this tower in Florida, in addition to damaging the house. (KE4AMW photo)

taken down and moved. Hey, I don't make this stuff up!

Let's review, shall we? A Cushcraft R-7000 vertical mounted on a piece of pipe pounded in the ground is personal property. Eighty feet of guyed Rohn 55G in 15 cubic yards of reinforced concrete is a dwelling extension. No claims adjuster will ever confuse the two.

One other thing. If you have towers and equipment on property you own or rent *away* from your primary residence, make sure your agent knows. Some homeowner's and renter's policies have significant coverage limitations for off-premises structures (towers and antennas) and personal property (radio gear, computers and the like).

ARRL Ham Radio Equipment Insurance

A very viable alternative to your homeowner's policy is the ARRL sponsored Ham Radio Equipment Insurance Plan. More details on this plan are available from the ARRL Web site (**www.arrl.org**).

This program provides protection from the *property* hazards we all face and the accompanying costs that can follow from loss or damage to your amateur station and mobile equipment by theft, accident, fire, flood, tornado and other natural disasters. Antennas, rotators and towers can be insured too. And, unlike most homeowner's policies, your equipment is covered if damaged by an earthquake or flood.

You're even covered for the replacement cost of computer software such as disks and tapes, including reimbursement of the expense of reprogramming for up to $1000 per claim, if you have computer hardware scheduled on your policy. Computer equipment may include televisions, recorders and other monitoring systems that are related ham radio accessories. And, your equipment and accessories are covered in your car, truck or RV, as well as your home.

You should compare other insurance coverage with the ARRL plan. You'll

find this plan less expensive with broader coverage than similar plans available. In fact, the ARRL "All-Risk" Ham Radio Equipment Insurance Plan is one of the most comprehensive policies you can buy.

ARRL Insurance vs Your Homeowner's Policy

Most homeowner's insurance policies will provide what is called "named peril" coverage for your accidental direct physical loss to your contents or personal property. The coverage can vary from policy to policy…but here is a partial list of named perils from State Farm's Ohio Homeowners Policy:

1) Fire and Lightning
2) Windstorm or Hail
3) Explosion
4) Riot or civil commotion
5) Aircraft
6) Vehicles
7) Smoke
8) Vandalism or malicious mischief
9) Theft (with limitations on specific property, such as jewelry and firearms)
10) Falling objects
11) Weight of ice, snow or sleet
12) Sudden and accidental discharge or overflow of water or steam from plumbing, heating, air conditioning or fire sprinklers
13) Freezing of plumbing, heating, air conditioning and household appliances
14) Sudden and accidental damage to electrical appliances and systems from variations in electrical current
15) Glass breakage.

Each of these perils comes with some limitations that are defined in your policy. For example, you can't go to Florida for the winter, turn off the heat in your home and then file a claim for frozen pipes. *Again, there may be significant variations from policy to policy, insurance company to insurance company and state to state. Make sure you contact your agent for clarification.*

Some homeowner's policies still provide named peril coverage for

Lightning can strike anywhere in the country, and tall towers can be vulnerable. (N3ZK photo)

dwelling extensions (towers and related equipment), but most cover accidental directly physical loss (an insurance policy term), again, with some exclusions. Typical exclusions would be damage resulting from flood, normal wear and tear and earthquake.

The ARRL's plan provides accidental directly physical loss coverage on all ham radio equipment, computers, towers and antennas. That means, unless the damage was caused by something specifically excluded in the policy, your claim will be paid.

One exclusion in the ARRL's plan is equipment damaged while you're repairing it. So if you're hard at work on the Loudenboomer 9000 amplifier with the power on and you drop a screwdriver in the power supply cage and it destroys the amp...sorry, no coverage. Another exclusion would be normal wear and tear. If your coaxial cable breaks down after being up on the tower for 15 years...sorry, no coverage.

Depending on your homeowners insurer, if you file a claim — even a small one — your policy may be surcharged for the loss. If you start having a higher than average number of claims (what people in the insurance business call a loss frequency problem) you may find yourself getting non-renewed and will probably find it difficult or impossible to purchase homeowner's insurance

from any carrier, at any price.

In Ohio, where I live, the average homeowner has a property claim that exceeds the policy deductible every 12 to 15 years, depending on the company providing the insurance and where the homeowner lives. If you turn in a homeowner's insurance claim every couple of years, sooner or later you *will* get a letter from your insurance company that won't make you or your mortgage company or, most importantly, your spouse, very happy.

Now if you're asking yourself, why should I consider the ARRL's plan when I already have a homeowner's policy, let's review:

1) Broader coverage on all ham radio gear than your homeowner's policy.
2) Lower deductibles than the typical homeowner's policy.
3) Claims activity won't affect your homeowner's policy.
4) Happy spouse...happy house.

Everybody's situation is different, but here I am, an insurance agent, telling you the ARRL's plan might be a pretty good idea for you to consider. Hey...if I were you, I'd listen to me. It could insure your domestic tranquility.

If You Rent

Hams who rent houses or apartments also need coverage. Some renter's pol-

icies provide personal property coverage only. Others provide 10% of the contents coverage amount for dwelling extensions, like a homeowner's policy. Again, see your agent for clarification.

If you don't have renter's insurance, get some. That's especially important if you have a tower, if for no other reason than to have liability coverage. Which brings us to the next big topic

Liability Insurance (Don't Forget Your Umbrella)

The first thing I learned in Insurance School is: either you either buy an insurance policy for those risks or items of personal property that you can't afford to lose (transferring the risk of loss) or you self-insure (assuming the risk of loss yourself).

People driving 1983 Yugos don't buy collision coverage. Those driving new Corvettes do. (Do the math.)

Like a swimming pool or trampoline, your tower is what attorneys call an *attractive nuisance*. If a neighborhood kid climbs your tower on a dare and falls...you're going to get sued... for a huge pile of cash...and you're probably going to lose. This is called bodily injury liability. If your tower falls and damages your neighbor's home, car or other property, his insurance company may try to recover the damage caused from you. This is, duh, property damage liability.

The good news is there's *some* protection in your homeowner's policy, typically *only* $100,000 to $300,000 of personal liability coverage. The insurance company will hire an attorney to defend you and write a check for damages up to the policy limits.

The bad news is when an attorney parades a teenager in a wheel chair in front of a jury, policy limits of $100,000 to $300,000 are just a down payment. (Your argument that this kid was *trespassing* will almost surely fall on the jury's deaf ears.) *You* are responsible for the rest of the damages, if you're collectible. You're collectible if you (*or your spouse*) have wages or other income, equity in your home or business, bank accounts, investments, retirement plans, future inheritances or any other assets that can be attached by court judgment.

If you can afford to lose these things, fine. If you can't, you need to purchase a personal liability umbrella policy (PLUP) with coverage of *at least* $1,000,000 per occurrence. It doesn't cost a lot, typically less than a dollar a day and will provide great piece of mind. There are many other benefits to owning a PLUP; see your agent for details.

What's the bottom line on justifying a PLUP purchase? If you're sued for everything you earn or own, now and *in the future* and if you *lose*…who do you want to write that check? You or your insurance company?

I thought so.

By the way, that check for everything you've got is the minimum amount of coverage your umbrella should provide. A little extra wouldn't hurt. Typically, personal umbrella policies are sold in million dollar increments.

Now…go back and read this section again. *It's very, very, very important.* After you've done that, get to a phone and call your Insurance Agent and tell him you need a new umbrella…today!

Here's a quick, but very important note. *The ARRL's Insurance Plan does not include any personal liability protection at all.* This is an area you will need to discuss and review with your insurance agent.

Filing an Insurance Claim

"Honey, did your tower always look like that?"

It might have been a tornado or other windstorm, lightning strike, ice or vandalism, but your tower and antennas are damaged. Aren't we feeling better that we discussed things with our agent before a Bad Thing happened? Now, it's time to get busy.

Your primary duty after a loss is to protect your property and the property of others from *further* damage or loss, then make reasonable and necessary temporary repairs. You obviously want to prevent additional damage, but your insurance company doesn't expect you to kill yourself doing it. It might also be a good idea to take some pictures.

Notify your agent of the loss as soon as possible. Many agents and companies have 24/7 claims service, so if the loss happened on Sunday morning make the call. It would be appropriate to follow up with a short letter, fax or e-mail to your agent, just to make sure the claim was filed correctly, requesting the claim number and the claim representative's name and phone number.

After filing the claim, your agent's involvement will probably be minimal, unless there's some issue you and the claim rep can't work out. When your claim is settled, an "attaboy" e-mail to the agent *and* the claim rep would be a welcome, appreciated and very unexpected surprise.

Organize Your Paperwork

Start a claim file, which should include:

1) Notes on conversations with your agent and claim representative, including dates and times.
2) Cost estimates on items to be repaired or replaced.
3) Any time you, your friends or family spend on repairs or clean-up. Your claim rep may "pay" you for this time by allowing you to offset it against your policy deductible. This is negotiable and should be discussed early in the process.
4) Photographs and inventory of damaged or destroyed items. Do not throw *anything* away until the claim rep says to. Damaged property that the insurance company pays you for becomes theirs and they may elect to sell it for salvage value.
5) Related claim data: the claim number, claim rep's name, postal mail and e-mail addresses, office, cell and fax numbers.

Get prices from several vendors. If the items destroyed are no longer available, most insurance policies provide for replacement with items of "like kind and quality." Document these items and discuss them with your claim rep.

Remember that your claim rep is probably way out of his or her element. Your damaged tower may be the first one they've ever seen, so they'll ask lots of questions and need some documentation. That's okay. Claim adjusters are trained to pay every penny the company owes — not a penny more, not a penny less. The claim rep will probably have to justify your settlement check to a supervisor. Make that job as easy as you can. Trust me when I tell you that the rep wants this claim resolved just as much as you want to be back on the air.

Also, as my sainted Grandmother used to say, "Pigs get fat, hogs get slaughtered." Some people think an insurance claim is like hitting the lottery. Not so, Bucko. If you can buy new gear at a great price, turn that price in — not the list price. Claims people get real cranky (and rightfully so) if they feel a claim is being "padded."

In this day of Web sites and search engines, trust but verify is easily accomplished. Your claim rep may not know a Force-12 C31XR from next Tuesday, but Google does. At best, padding slows the process and might leave you with egg on your face and some explaining to do. At worst, you may find yourself facing felony insurance fraud charges. A word to the wise should be sufficient.

One other thing to remember. Your damaged tower is not as high on the claim rep's priority list as someone's home with major structural or fire damage, nor should it be. When bad things happen, there is never enough time and people to get things done as quickly as people would like. Be patient and be reasonable.

For hams in the northern climates, ice and snow pose a significant challenge — not only to the tower, but to surrounding trees that can fall into the tower or guy wires. (KA1ZAD photo)

Making the Repairs

Most insurance companies write a check for the total damages (less your deductible) when the work is completed and all damaged/destroyed items are replaced. If you want a check prior to the start of work, the claim rep may reduce the payment for depreciation of the damaged property, paying the balance when the damaged items are actually purchased or work is completed. This is consistent with contract language in most homeowner's policies and keeps everybody honest. Also, don't be surprised if your contractor asks you to pay your deductible directly to him before he starts work. Pay by check, get a receipt and tell your claim rep.

Most homeowner's policies settle losses on a *replacement cost* basis, meaning you get new stuff for old, *when* the new stuff is purchased. Make sure you have replacement cost coverage on your dwelling, dwelling extensions *and* your personal property before you have a loss. This is another good reason to see your agent real soon.

If you hire a contractor to repair or replace your tower, *insist* he provide you with a current certificate of contractor's liability insurance (his malpractice policy), a current certificate of worker's compensation coverage and local references. If he's a professional, this won't be a problem. If he's a fly-by-nighter, he's probably not insured, and you just don't need those headaches. If the contractor tells you his employees are "subcontractors," you'll need to see the subcontractors' insurance documentation.

Do not even consider bending on this. Ever!

Your contractor may ask for some payment up front. It would be wise to include your claim rep in that discussion since he'll be writing the checks.

Seek Professional Advice

Here's the bottom line: Don't rely on your ham radio buddies, your neighbor or your brother-in-law for advice on insurance matters. Call your agent right now, update him on your installation and ask hard questions. "If my tower and antennas are damaged, how will my policy cover the loss?" If your agent doesn't know, make sure he finds out and then shows you in writing (in the policy) how losses will be covered or why they won't be. If there are gaps in coverage, what will it cost for what you need? Photograph your installation and give your agent an accurate inventory of your gear. Get to know your agent — after all, on the worst day you might ever have, he'll be standing beside you.

Here's another little secret…every time your pay your premium, your agent gets paid. (All right, I *said* it was a little secret.) If your agent can't or won't provide you with the service you deserve, he's not earning his keep. It's time to go shopping for a new company or a new agent with your old company.

Every insurance agent worth his salt has some war stories to share. Here's one of mine. Not long after my article was published in the February 2009 *QST*, I got an e-mail from a ham in North Carolina. It seems that he followed my suggestions, spoke with his agent and upgraded his homeowner's insurance policy for his tower installation.

Everything was just peachy.

A couple of months later, he got a letter from his agent advising him that due to his tower and antenna system, his current carrier was non-renewing his homeowner's policy. The ham sent me a photo of his home and his tower installation. I could see absolutely no reason for the company's decision. In fact, much to my chagrin, this guy did a better job securing and climb-proofing his tower than I did for my tower.

I referred this gentleman to an acquaintance of mine, who is an agent for a large Good Neighborly Insurance Company with an office near his home. A new homeowner's policy was written with that un-named insurance company that very day. The moral of this story is: There are hundreds of thousands of hams who are homeowners, with towers, beams and other antennas, who have no problem getting and keeping homeowner's insurance coverage. If my good friends Tim Duffy, K3LR, and Tom Lee, K8AZ can get a homeowner's policy for their homes and antenna plantations, believe me, you can too.

Finally, and for the last time, *please*

make it a priority to visit your agent and review your policies. Insurance agents are just like you. We don't like surprises and we don't like conflict. We do feel really good when we take care of a client's claim and put him back to where he was before the loss happened. As my teenage son would say, "it's, like, our job."

Ray Fallen, ND8L, has been an agent for The State Farm Companies since February, 1988. The opinions expressed in this chapter are solely his and are not necessarily those of the State Farm Insurance Companies. The coverages described in this chap-

ter may not be available or apply in your state, province or country. You are strongly encouraged to review your homeowner's policy and tower installation with your insurance agent to determine appropriate coverages and coverage amounts. The words "he/his/him/guy/attaboy" are used in a gender-neutral basis.

Ray's been licensed since 1964, an Extra Class Licensee since 1983 and a member of the North Coast Contesters and the K8AZ multiop team. Ray's a confirmed appliance operator, contester and DXer and has earned 5BDXCC and DXCC Honor Roll (Mixed).

13

Working with Professionals

You might argue that this is *amateur* radio, so you should be able to prepare the site, build your tower, install the antennas and maintain everything yourself. Or if not by yourself, that you could get the work done by simply using the resources found within your local ham club, or with help from other ham friends. And these truly are options, viable and real, used by countless hams to install and maintain their antennas.

They are, however, not always available options, nor always the best options. Sometimes the best and safest way to do the job is by using professional help.

Climbing and working aloft requires special tools. Building a tower is sheer physical labor — something some ever-aging (and often out-of-shape) hams are not capable of doing. At least not safely. And that's the single, largest factor you should consider — doing this type of work *safely*! That requires not only special tools, along with some solid physical conditioning, but also practical, very realistic levels of experience.

Experience is, of course, the basis of all knowledge, wisdom, understanding and meaning. There's simply no substitute for it. When your life hangs in the balance, experience is what you need to successfully complete the task.

So hiring an experienced worker to build, climb or repair your tower makes sense, in many, many ways.

> **"** *Choose someone who will complete the job quickly, efficiently, correctly and safely.* **"**

Conventional wisdom posits that the less money you pay for something, the better you'll feel about it. In my experience, the conventional wisdom is wrong, at least when it comes to tower work. You get what you pay for.

How do you put a price on avoiding an accident? How you know your installer's proposed design for your new tower meets your needs? How do you know his idea or solution for putting your big Yagi up in the air is right? How can you decide what's fair?

The answers to such questions cannot, and will not, apply to each and every situation. Every installation is different, even if it's just one small, seemingly insignificant way. So, experience and some history — with the type of tower, the type or manufacturer of antenna, the brand of rotator — all count for a lot. You're hiring someone to provide a specialized service, not shopping price on a box of pencils.

Buyers should choose their vendor or supplier based on the one who offers them not only a fair price, but someone they think will be able to complete their work quickly, efficiently, correctly and safely. While this is more-or-less common sense, it's also the basis of the legal aspect of what constitutes a bargain (a mutually-agreed upon arrangement between buyer and seller) instead of the more common layman's language — something that's merely "a good deal."

The *Right* Experience Counts

NBC's production of an episode of the TV show *Dateline*, entitled "Tower Dogs," generated a lot of traffic and conversation among hams. And probably more conversation spun up around a thrown-away line referring to the worker's $14 per hour pay rate than any other aspect that was presented on the program.

Within the past few years, I've more-or-less made my living by doing ham radio tower and antenna work. (No one is more amazed by this than I am!) And in that time, I've made a point of meeting, talking with and even hiring some professional tower guys to help me with various projects in various locales.

In every instance, the professionals wanted to know how I got into the business of doing this sort of tower work because the rate was better than what they were being paid. In every instance, none of them truly understood what a Yagi (or any other ham antenna) was, why and how we use rotators, how to solder, what HF radio means to us, or many other topics that we hams all take for granted. None of this surprised me.

On the other hand, in every instance, they had far more experience than I do with large Heliax feed lines, with cranes, and with working on really big towers and big hardware. Being younger guys, they climbed a lot faster.

Even though they may have gotten to the top of the tower faster than I did, I ended up teaching these non-ham professionals a lot about the work we needed to do up there. That's simply because they had virtually *no* ham-related experience and were unfamiliar with the types of antennas and rotators we needed to work with.

And this remains a critical factor, in my opinion. Clients should consider the experience level of any potential vendor or worker. You want someone who's experienced; that goes almost without saying. But how do you determine that they have enough of the right experience? In other words, if you feel you need to hire professional help, how can you judge the qualifications that vendor presents to you?

Finding and Evaluating Professional Help

In the world of tower dogs, there are over 500 member companies located throughout the United States, Puerto Rico, Bahamas, Canada and Cayman Islands. That's within the area served by NATE (the National Association of Tower Erectors), totaling about 95,000 individual workers involved in the industry.

In the world of ham radio, as this is written there are fewer than two dozen individuals or companies providing antenna and tower services represented on the review page at **www.eham.net**.

And of those, only a handful have had recent reviews. That's a staggeringly small number, considering we've got over 650,000 licensees spanning all areas of the country. Sure, not everyone is active or will need or want or can even *have* a tower, but still, the resource pool of qualified tower personnel is very limited.

> ❝ *Safe tower workers always think about where they are.* ❞

Given the safety concerns (both during the initial installation and over the life of the tower and antennas), and given your substantial investment in time and money, it only makes sense to hire the right contractor for the job. This requires you, as the customer, to be more proactive in choosing someone to do such work, in order to make a meaningful evaluation. You can gauge that level of experience by reading those **www.eham.net** reviews, for example. You can simply ask whether or not the vendor has done work like yours before, and so forth. Most vendors will supply you with references, if you ask. Most vendors will have the needed tools, so the experience level is the only real variable left for you to consider.

Again, safety should always be your main concern. It's been said often enough that it does not require real smarts or bravery to climb a tower. It does take trust — trust in your equipment and your abilities. Climbing experience doesn't mean someone isn't afraid of heights, either. (My climbing partner John Crovelli, W2GD, knows he can always make me smile by saying, at some opportune moment, usually when we're above 100 feet: "I'm afraid of heights.") Safe tower workers always think about where they are (whether that's 20 feet or 200 feet up a tower), how they got to that level, and how they're going to get *down* from that level. They always respect the fact that a mistake and fall from any height can cause serious injury or death.

This mostly all means we're comparing apples to oranges if we compare the Tower Dog's world to what we do. We can learn things from them, of course (anytime you encounter guys with that much experience, you should try to benefit from it). But if you saw the show, you may have noticed that nearly every single structure they worked on was unique. That's a major difference. Much of what we do is confined to similar towers, using similar antennas, using similar hardware — a *lot* of the same stuff, over and over, again and again.

Much of what we do is not governed by deadlines — we can work at our own pace, regardless, weather permitting. Much of what we do is not driven by a business plan — it's only a hobby, after all. And what we do is not regulated — there's no OSHA stipulating what equipment we use. But everything said about safety certainly applies. Always.

Ham versus Commercial Towers: A Professional's Viewpoint

As is often the case, there is a "professional" side to tower work and installations. Indeed, commercial towers often dwarf ham installations. And while one's first reaction to these towering beauties might be to ignore them, it makes more sense to me to study them. And to analyze and inquire about them. Indeed, I decided to ask someone who's been in that business for a number of years if it would be possible to compare and contrast the two fields — thinking it would be profitable to learn from the pros. After all, it's probably more than a simple question of scale when it comes to comparing commercial installations to what you have in your backyard.

So, I went to Norm Jeweler, W4NRS, who owns United States Tower Services, in Frederick, Maryland, which has been in business for 40 years. USTS has never had a fatality, never had a tower fall or collapse, and is a leader in the tower industry's safety program.

I asked Norm a few basic questions. Here are his answers.

Q. Take a moment to compare and contrast a "typical" commercial tower installation to a "typical" ham radio installation.

A. There is really not much difference between a ham tower installation and a commercial tower installation. Commercial towers are just usually bigger and taller with bigger antennas on them. A tower is a tower is a tower! But, they all require knowledge of the product and how to install it safely without getting hurt, so the owner ends up with a quality installation that will last throughout the years. A commercial tower installation usually lasts 30 to 50 years with the right kind of maintenance.

Q. What's the most obvious or common mistake(s) you've seen or encountered over the years with that typical ham's tower installation?

A. The biggest mistakes I see in ham tower installations are not using the correct safety equipment — like an approved climbing harness. Other problems include using the wrong equipment to raise or lower tower sections and antennas (gin pole), not wearing a hard hat, free climbing or working by yourself without someone on the ground to be there if you need help. Working without a safety plan before you start the project is another thing many hams forget or don't do. Or not checking the area for obstacles such as trees, power lines, roof lines and anything else that will be in the way while installing a tower or antenna.

And, of course, not following the manufacturer's directions, that's always a factor. When a manufacturer says the base of the tower should be 3 × 3 × 3 feet with one yard of reinforced concrete, there is a reason. Don't underestimate their knowledge and use less. Or when ½-inch reinforcing bars are called for and you use wire mesh to hold the concrete together. Things like that.

And, I should say to always finish the concrete at the base of a tower so water

will run away from the legs and not pool up around them. When installing hollow legged towers, be sure to put the legs in at least six inches of gravel to allow for drainage of condensation. Never step on the base section and push it through the gravel, which plugs the legs with dirt.

And I can't forget, using water pipe or fence pipe for mast. Bracketing a tower to the side of a house or building without reinforcing the area on the other side of the bracket. Hams often overlook things like that to try and save money, but it's false economy.

Q. What steps should hams take in order to work as safely as possible on their own tower installations?

A. Again, the most important things are using an approved harness (not some used leather climbing belt that's dry rotted). The only truly safe belt is a nylon web sit harness style belt with approved safety straps (yes, at least two). Then, stay away from overhead electrical lines, wear a hard hat, make sure you have a ground observer who can call 911 if you get into trouble. Use only approved equipment. The days of using 2 × 4 lumber or conduit for gin poles are hopefully long gone. Use a nylon or Dacron rope on a tower job, not manila or poly ropes.

Q. Since it is, after all, "amateur" radio, should a ham who wants a tower ever consider hiring someone to do the work — such as setting the tower and guys, building antennas and then installing them, and so forth? Or should such work be done by the owner, perhaps with the help of local ham buddies or some radio club pals? In other words, is there a place for tower "professionals" within the ham radio ranks?

A. In my estimation, the answer to this question is *absolutely*. If one doesn't feel comfortable climbing or doing all the labor to dig the holes and installing the concrete for a tower, then simply leave that to a professional. You can easily overstress yourself by digging the base or anchor holes and die

from a heart attack. Pushing yourself to climb a tower if you're not 110% comfortable doing so means you could fall. One must ask: Is paying a professional to do a guaranteed job worth my life and limb? Will I sleep better at night knowing that such work was done properly and will last for 30 or more years?

Towers installed by professionals (by professional I mean someone in the business at least 15 years or more, who belongs to the National Association of Tower Erectors) will provide that peace of mind. And, if you want this professional to install your antennas and rotators, get someone who has done this many times before — not just once or twice, and who knows what they're doing. You'd be surprised at the number of professional tower company owners and climbers who are knowledgeable hams.

I'm 64 years old and I have an 80-foot free standing tower at my Florida home. I know better than to climb my tower anymore. I'm overweight, have some health issues, and I want to live to enjoy my hobby for many more years. So sure, I gladly negotiate with my local professional tower company to do work when I need it. It's very reasonable.

Remember: if you ask friends or members of your local club to come over and help you on your property to work on your tower and they get hurt, guess who pays for their injuries, or heaven forbid their death. One more thing. Your homeowner's insurance likely will not represent you or pay a dime in your defense, much less pay a claim for something like this.

Q. If one already has a tower, talk a bit about maintenance — what it means and might include.

A. Maintenance is the key to having a tower that's safe, reliable, and will survive winds and ice and last for 30 or more years. At minimum, guyed towers should be inspected at least twice a year, including checking guy tensions, checking legs for splitting at the base due to water in them, verticality, and any loose or missing hardware. Grounding should also be inspected twice a year, too, and this includes checking the connection to the ground rod.

Q&A With Professional Tower Installers

What began as a couple of rather simple questions from the editor working on my manuscript for the ARRL quickly morphed into a larger group of questions, which further fueled the idea of asking a number of tower installers for their opinions!

So, I polled several of my fellow tower climbers who have a lot of positive reviews on **www.eham.net**, and all agreed that the idea had merit. Everyone responded within a few days of one another.

While these are all East Coast workers, I don't believe there's too much difference between working hereabouts and working in southern California, let's say, except for the weather in mid-January….

So, here are six general questions and the answers from not only me, but also from John Crovelli, W2GD, Dan Street, K1TO, and Ray Higgins, W2RE. I've provided everyone's relevant contact information, as well.

Tower Works
Don Daso, K4ZA
515 Withershinn Drive
Charlotte NC 28262
Phone: 704-408-7948
Email: **k4za@juno.com**

John Crovelli, W2GD
PO Box 10
Solebury PA 18963-0010
Phone: 908-391-5611
Email: **w2gd@hotmail.com**

A1 Tower Service
Dan Street, K1TO
9933 289th Street East
Myakka City FL 34251
Phone: 941-780-0073
Email: **A1TowerService@aol.com**

Hudson Valley Towers
Ray Higgins, W2RE
www.hudsonvalleytowers.com
Phone: 888-528-6937
Fax: 866-509-0843

1) Do you work with clients who are just starting out and have absolutely nothing? (That is, would you do an entire project from concept to connecting the last cable?)

K4ZA: I've done numerous projects for "first time" tower owners! I enjoy the teaching aspect of explaining and laying out the various options for clients. In fact, I truly like planning, and then building, a station. So, sure, I've started with a drawing and worked through the process, right up to handing a newly-finished hunk of coax to a client, saying, "Here you go…have fun!"

W2GD: I work with clients ranging from newbies to contest or DXing fanatics, providing consultative services at whatever level is requested or felt to be necessary to move a project from concept to completion.

K1TO: Yes. Site and materials planning is often very helpful to them, since there are a number of factors they typically are not thinking of initially.

W2RE: We all remember what it is like when starting out with anything new. It can be trying to find people to help and "Elmer" you along the way. We try to keep an open door policy for newcomers in the hobby, and even though they might not have the budget to hire professionals, we feel a few phone calls and emails pushing them in the right direction always seems to be the right thing. We understand there's a lot to learn from this great hobby and we do our best to help even if monetary compensation is not an option. So really, we have done small jobs — from the new ham who just got his ticket and we installed a vertical on his chimney — to full out soup-to-nuts installs with stacks.

2) Along those lines, if the client needs professional engineering drawings for local building officials, do you give references? Or do you only come in after the planning or permits are done?

K4ZA: I've worked with a variety of professional engineers (PEs) for clients who needed that. My favorite

is Hank Lonberg, KR7X, who contributed information on freestanding versus guyed towers and star guys versus normal guys for other chapters of this book. Being licensed, and having experience with contesting and big stations, he's truly a valuable resource.

W2GD: In general, I don't become directly involved with zoning or permits. However, it's certainly in my best interest to explain the process to a prospective client. I guide the client on what to do, with whom, and when. In some cases I point them toward local government Web sites, where they can research their zoning laws. Other times, I may direct them to a PE for assistance, and in the most difficult cases, to specialized legal assistance.

K1TO: I prefer not getting involved in local permitting. I always use referrals for professional engineering drawings.

W2RE: We try to give references as much as possible, being proactive and helping with tower specs and referring legal council when dealing with zoning boards.

3) Do you charge clients by the hour or by the job? And do you have different rates for different work? (For instance, building a rebar cage and installing an 80 meter beam at 200 feet would seem to command different rates, but maybe not.)

K4ZA: I work by the hour. With hams, there are too many variables, and any other way is too risky! The level of knowledge needed to do the work you describe does not relate directly to the labor required in your examples. Wire-ties on some rebar sounds ridiculously simple, while putting up an 80 meter Yagi sounds terribly complex. But there's lots of information and knowledge needed to correctly accomplish each of those projects, for instance. I'm not adverse to discounting work, though. Crank-up installations, which sometimes don't require much climbing, often come in cheaper than full-blown, guyed tower jobs, where I'm spending a lot of time climbing and hanging on the tower.

W2GD: In almost all cases, my work is done on a per hour basis, at the same hourly rate regardless of task performed. Some discounting is given, at my discretion, when I feel the work doesn't support the amount charged. I have a four hour minimum charge, which is sometimes waived if the job takes far less time. Fixed pricing is too risky.

K1TO: Typically, by the hour, usually with an up-front estimate so we are both "calibrated." But a few jobs have been "fixed bid" projects. And we definitely charge less for ground work.

W2RE: It all depends on the scope of work — if the quote is either hourly or one set price to complete. Sometimes it's very difficult to quote out larger jobs because of unforeseen issues that may arise. Another factor that affects a rate is level of difficulty, or if any special precautions need to be followed. Usually, on every request for work, we spend some time gathering as much information as possible beforehand (pictures, either Google Earth or site inspection) to get a good handle on what needs to be done and what obstacles may be encountered.

4) How do you decide how many paid helpers are needed (and the client pays for)?

K4ZA: If the workload is large enough — say we're putting up a 40 meter beam, I'll bring a helper. Or quite often these days, as our population is getting older, my client cannot help out directly, for health or physical reasons, so then I'll bring along a helper. That's fairly common. That's all discussed beforehand, so there are no surprises.

W2GD: The work dictates how much manpower is needed to complete the project. Typically, no more than one helper is needed. It's the rigger's decision, not the client's...always!

K1TO: If the client is competent and willing, we have no problem working with them directly in lieu of bringing our own personnel. Scope of the project dictates personnel needs. On occasion,

I've used fewer resources for a longer time to accomplish the same result.

W2RE: Once we've determined how we're going to complete the request, the rate given will include everything and everybody on the job. If we feel that at some point we may need a extra hand or two, there would be no extra charge to the customer. The price for the whole job or by the hour will not change depending on if we bring more people in to complete it. This way, the client doesn't have to worry about extra cost at job completion. At no time do we want the client to feel there may be unknown charges that have not been discussed ahead of time.

5) If the client or their friends want to help on the ground, is that okay? Encouraged? What if the client wants to climb and work with you?

K4ZA: Sure! I've actually "taught" some clients how to climb. Bought them their fall arrest system and everything. But not many clients climb. I certainly don't mind if the client wants to work. I think that's great. Frankly, I'm surprised when they don't get involved — wanting only a turnkey job. But that's just me. I'm a nuts and bolts person, who likes making stuff. So I think everyone should be excited about building a beam in the backyard! But then I remember everybody's different, and that's why they hired me in the first place! It *is* very, very common for me to be the climber, and for the client to be the ground crew. A lot of clients simply don't, won't, or can't climb their tower. But they still do everything else. Whatever it takes…I'm a blue-collar guy. Work is work.

W2GD: I have no problems with a client climbing. I don't generally encourage this (frankly it's quite rare), unless there's some significant advantage to having two on the tower to do a given task. Of course, you must use good judgment, and make the decision on a case-by-case basis.

K1TO: Again, if the client is competent and willing, we have no problem working with them directly in lieu of

bringing our own personnel. But typically, if the client climbs, he/she isn't in the market for tower services. Many clients are ex-climbers and thus are quite competent on the ground.

W2RE: This isn't encouraged and we very *rarely* allow it. First and foremost is safety. Things can happen very quickly and unexpectedly and the last thing we need to worry about is the client being in the wrong place at the wrong time. The ultimate goal is for the job to be completed above the client's expectation and most of all, without injury or damage to anyone or property. It's just like hiring any type of professional — like a plumber or an electrician. It's best to leave the work to the professionals.

6) Who arranges for local equipment, such as a backhoe, the concrete, crane and so on?

K4ZA: Good question. Some locales are so small that there isn't a Sunbelt or an RSC or other national chain nearby, so the client will not only have to supply the name or source, but do the actual rental, too. I have accounts at both places, but not at Joe's Tool Supply, you know? So I'll just tell the client what we need, what to get, and so forth.

W2GD: Since I don't generally dig in the dirt (since it's not cost effective for my clients), foundation work is almost always left to the client, which includes arranging for a local general contractor, equipment and materials. I've never arranged for a manlift or crane; again, these items are ordered and paid for by the client (although I will help them make equipment selections).

K1TO: I prefer the client line up local resources, especially for out-of-town jobs. This gives them the opportunity to use a preferred resource and to shop around if desired. This is usually how it works out, but again, on occasion, we'll have to line up everything remotely.

W2RE: We try to handle as much of the job as possible. As a matter of fact, we encourage letting us take full control of the project. This allows us to coordinate all parties involved and have a solid timeline of all the parts of the project. If too many people are trying to coordinate the various vendors needed for a job, then the risk of miscommunication and error will occur. The worst part of this is loss of time and productivity. Nevertheless, if the client wants to handle some part of the job — for instance, taking care of the base, concrete and getting the first few sections in — we have no problems with that as long as the client follows the manufacturer's specifications. We also ask for digital pictures before hand to make sure nothing was missed and that the structure is safe and secure before we show up to finish.

Tools and Gadgets

Sometimes problems demand unique solutions, which is where Plato's observation that "necessity is the mother of invention" probably came from. Tower work isn't any different.

A Fixture for Crank-Up Towers

The first tool here was suggested to meet a need during an installation for Rick Kourey, K4KL, near Charlotte. Rick had selected an 89-foot ITS crank-up tower from Array Solutions, along with an OptiBeam Yagi for 20/17/15/12/10 meters and M² antennas for 30/40 meters and VHF. Some photos of Rick's installation are shown in other chapters of this book.

The approach was fairly straightforward, involving a Genie 50-foot man-lift to install a 3-inch high-strength mast, along with the topmost 2 meter vertical and the 30/40 meter and 6 meter Yagis. But, by then, we'd run out of daylight. We had, how-ever, also carried the OptiBeam into position next to the tower. All hands pressed for a quick raising using the good old gin pole, but I wanted to ensure the beam itself was well away from the tower base. It was pretty crowded there with the crank-up's motor, the control box, the tower grounding items and so forth. I opted to construct the simple cantilever rig described here, and install the 5-band Yagi by hand on another day.

Selecting some 1-inch square tubing, the main arm was welded up, using two sections; then, suitable braces were cut and welded to it, using the same stock. These were then bolted to a plate of ¼-inch thick 6061 alloy aluminum. All materials came from the K4ZA collection of metal "I knew I'd find a use for one day," which seems to grow exponentially as time goes by.

Figure A-1 shows this simple device's construction. The ability to keep the beam away from the motor and hardware near the ground makes raising the beam on crank-ups or on big taper self supporting towers an easy task. Once at the top of the tower, I simply attach a come-along (mounted to the mast, this time) and swing the beam into its final position.

Levels and Measuring Tapes

Levels are useful to any tower builder, with the 48-inch carpenter's level being the standard. Pocket levels are sometimes useful, even on the tower, just to give yourself some advantage over eye-balling things.

Not many folks pay attention to the angle of their guy anchors,

Figure A-1 — This homebrew yardarm or standoff tool is useful with crank-up towers to haul materials up outside the large base profile.

even though Rohn specifies the angle relative to the ground. Greenlee makes a truly handy level, with a magnetic base, that makes setting the anchors and aligning masts easier. Designed for electrical contractors (who must often bend conduit, usually at 30, 45 and 90° angles), this lightweight level now always goes on certain tower jobs.

Another useful level is what's known as a Post Level. It's the ideal tool for leveling ground poles, roof mounts, wall mounts or even fence or deck posts, which makes it ideal for radio masts, too. Most models are magnetic, and it's the only level allowing you to check for plumb without moving the level. It's perfect for installing tower bases, elevated guy posts, and the like. Check your local hardware emporium or Amazon for some great deals.

Hansom has come out with a new level design that uses a ball instead of the more common bubble (**Figure A-2**). The design makes reading plumb or level, as well as angles, easier than the old-fashioned bubble designs. It seems to make tower plumbing much easier, for some reason!

The Wixey Gauge (**Figure A-3**) is a clever and handy woodshop tool normally used to set saw blade angles. It's perfect for measuring the taper of self-supporting tower base legs.

Measuring tapes are usually taken for granted, unless you build lots of stuff. Stanley's latest Fat Max line offers some unique approaches. Advertised to stand out unsupported to 11 feet, mine only goes a tad past 10 feet — still way past any other tape I've owned. It makes single-handed construction much easier. The bulky case might be a problem for someone with small hands, and the black, rubberized case coating can get awfully hot if left out in the sun. And the spring return is *rapid*, and the curvature of the blade (a contributing factor to that superior standout length) can slice your finger quite easily. They're a bit pricey, too, compared to other tapes. But every time I actually use mine, I admire most of the pros and decide the cons aren't so bad after all.

Project Calculator

Every now and then, something comes your way that's so simple, yet refined, so inexpensive (*okay, cheap*) yet versatile, and so impresses you that you simply have to have it.

The *ProjectCalc Plus* pocket calculator shown in **Figure A-4** has to be one of the neatest tools I've found in the past few years. Yes, it's yet another example of computers being found in everything, from your toaster to the family car. It's so cool that you quickly forget that part, and look around for clever and cool problems to solve. And while there are several such calculators on the market, this one impressed me first with its price.

If you need to figure out:

- Paint coverage (surface area, or gallons, quarts or pints)
- Wallpaper (rolls of)
- Tiles (the number needed)
- Concrete (number of bags or mixer loads needed, for tower bases or driveways or fences)
- Gravel (shovels or truckloads)
- Bricks or blocks (number for a wall, driveway, whatever)
- Carpet (length or surface area)
- Fences (number of posts and rails, boards, wire)
- Cost (using custom or dedicated settings)
- Conversions (for the metrically-challenged or non-US users; it works both ways)

Well…you get the idea. This is one handy gadget! It's the size of your typical pocket calculator — meaning it will, literally, fit easily into your shirt pocket. The instructions are printed (in that annoyingly small type) on one sheet of multi-folded paper, which is

Figure A-2 — The Hansom ball level is precise in any direction.

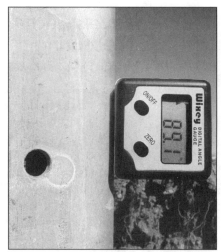

Figure A-3 — The Wixey gauge is a handy tool for measuring the taper of self-supporting towers. (AB7E photo)

Figure A-4 — The low-cost ProjectCalc is available from big box home stores, as well as online sources. It's one unique, versatile and fun-to-use tool.

bothersome at first, but then you discover that using the ProjectCalc is fairly intuitive. And, the paper folds small enough and is thin enough, that it fits inside the fold over lid or cover. You can take it out, store it safely in your shirt pocket, do all the functions (probably without ever referring to it), and then put it back. But it's always nice to know you *can* look something up should it be necessary.

It's simple and easy to use the calculator. You enter things just as you'd say them aloud, and suddenly, there's your answer, in the same format. For instance, if you want to add 3 feet 4½ inches and 5 feet 9¼ inches, you simply type

3 FEET 4 INCHES ½ + 5 FEET 9 INCHES ¼ =

and then

9 FEET 1¾ INCHES

magically appears on the screen.

Want to describe your tower accurately in your next DX QSO? Type

100 FEET CONV METERS

and

30.48 METERS

pops right up on screen.

These are but two very simple examples. The calculator functions are capable of holding specific or dedicated amounts in memory, along with standard-of-the-industry default values for bricks, block, concrete, plywood and so forth. For instance, it will figure cost per bag, paint coverage, size, board or post centers, studs, settings per tile or block and so forth. It will figure board or post on-center spacing (the actual board or post width plus spacing between them).

The ProjectCalc is obviously designed for the Do-It-Yourself market, where the math isn't so much hard as it is tedious and atypical. Never again will I find myself re-checking the math before calling the concrete supplier for a tower base pour. (I'm sure I'll still measure twice, but won't be multiplying the dimensions more than once.)

Moving Heavy Objects

I sometimes find myself faced with taking down crank-up towers. Then, I'm faced with moving them.

These are always heavy, awkward towers, and they don't easily come apart into manageable sections. Moving them requires planning and preparation. Part of that prep work is bringing along my tower dollies. One is a simple "wagon style" dolly, with ten-inch pneumatic tires, capable of supporting 1200 pounds. I have a 10-foot handle for it, which allows me to walk and steer with a collapsed tower hanging over the front. The other dolly is a modified "trailer hitch" dolly, which was intended to allow one person to roll a trailer around. I welded some plate to the hitch point, and this dolly now slides under heavy stuff and lets me roll it around. This dolly usually brings up the rear end of whatever long, heavy object I am maneuvering around some remote place. And I can still screw a ball on to it, and then use this dolly as originally intended. Together, these small dollies allow me to haul crank ups around relatively easily. I purchased the wagon dolly from TEK Supply and the trailer dolly from Northern Tools.

For moving heavy objects short distances, it's hard to beat a Lug-All lever ratchet hoist (which nearly every-one calls a *comealong*) shown in **Figure A-5**. The very first time I used one, I knew it was the best comealong in the world. Compared to the usual hardware store version, they're much better made and easier to use (they don't snag). I now have three of them.

If you think you're interested in moving heavy things around without hurting yourself, or spending large amounts of money to rent mechanized equipment, you might enjoy what one enterprising retired carpenter in Michigan has been able to accomplish. Visit **www.theforgottentechnology.com** for a truly fascinating look at some old methods and techniques for moving heavy stuff around by hand!

Sometimes, a professional tool or gadget gets replicated and offered for sale to the general public. That's the case with the Accu-Chute wheelbarrow. This heavy duty wheelbarrow was designed from the ground up to be used to transport concrete — with its heavier-than-usual base, large pneumatic tire, and a tapered chute (pouring spout) on its nose. Now, you can walk into a home center and buy a knockoff that'll do the same thing at half the Accu-Chute's price tag. This design is perfect for moving concrete and pouring it into tower base holes, anchor holes, and so forth. (Be prepared, however — when full, the wheelbarrow will weigh over 200 pounds.)

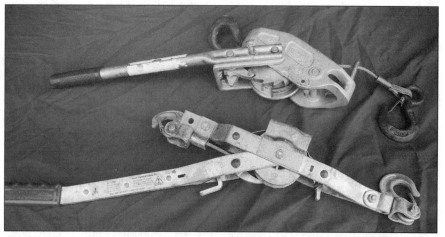

Figure A-5 — The Lug-All comealong (top) works more smoothly and is stronger than the hardware store version.

Wrenches and Pliers

Folks sometimes ask me about brands of tools, specifically wrenches. Here's my opinion: Snap-On makes the best wrenches, period. They are also the most expensive. And you must track down the Snap-On salesman servicing your area — that's the rolling display truck you can spot at auto dealerships and other professional service areas. That truck can supply everything you'll ever need. You can order online, as well, but I urge you to take a peek inside one of the trucks. You will be amazed!

Mac, Matco and Cornwell are other professional brands — also available from similar display trucks. Craftsman tools, available at Sears, will suffice for the average ham, shade tree mechanic or homeowner. Their lifetime warranty makes them attractive, as do their sale prices. SK tools are another brand worth your consideration.

Socket wrenches come in 6-point and 12-point styles. The 12-point wrenches are more useful for general work, but 6-point wrenches have more gripping power. If I had to choose one style over another, I would pick 12-point wrenches, although I've got complete sets of both in my toolbox. I normally only carry "deep" sockets up the tower, simply because they're more versatile, when working with U-bolts and the like. Unless I'm working on really big hardware, a ⅜-inch ratchet

Figure A-8 — A ViseGrip chain grip takes a good bite on masts or booms.

drive should work fine. (Again, I have ratchets and socket sets ranging from ¼-inch up to ¾-inch-drives, but I do more of this work than the average ham, so don't think you have to have everything in your toolbox that I do.) Occasionally you may need socket wrench accessories such as extensions, swivels or breaker bars.

One of my more interesting sets of tools remains my ratcheting box/open-end combination wrenches. What's unique about them (besides the novelty of a ratcheting box-end wrench) is that the ratchet works in only 5° of rotation. When you're working atop a tower, sometimes in cramped conditions, that smaller ratchet angle can make a big difference and save you time. Plus, the box-end wrench gives you a little more torque compared to a typical socket wrench. GearWrench, Craftsman, Husky, Snap-On and others all market these wrenches. Mine have proven so useful that I now have two sets, so even the ground crew can benefit from them.

GearWrench (and probably others, by the time you read this) now have some variations on this design. You can get a flex-headed set (great for tight spaces) and an X-box design, which allows you to put more force on the bolt or nut. **Figures A-6** and **A-7** show some of the hand tools I use.

For years, I relied on Klein lineman's pliers to solve lots of gripping and cutting problems. Recently, I have found the Craftsman Robo-Grip pliers to be more useful when faced with aluminum elements, larger connectors on baluns,

Preformed guy grips, and so forth. They can make the difference when taking down some old beams, for instance — providing a solid grip on hardware that's so rusted regular wrenches won't fit. Several manufacturers now make similar pliers, along with Craftsman. I'm carrying a little more weight in the pouch, but again, they can save time.

Figure A-8 shows the specialized tool of choice if you need to rotate a boom or mast while in the air. The ViseGrip chain grip takes a good bite on tubing without damaging it.

Other Handy Tools

I used to carry multiple screwdrivers with me up the tower. But a few years ago, I switched to one of those cheap multi-bit screwdrivers — two sizes

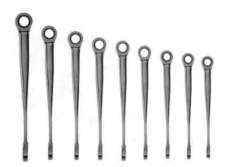

Figure A-6 — GearWrench ratcheting-gear end wrenches.

Figure A-7 — A typical set of deep sockets.

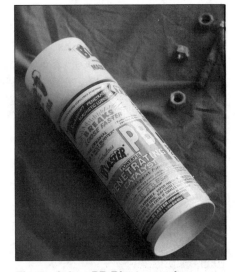

Figure A-9 — PB Blaster works wonders on rusty hardware.

of straight blades and two sizes of Phillips blades. Only once during those two years did I need a screwdriver that would provide a bit more torque. So now my pouch is lighter by not carrying so many screwdrivers. I liked this idea so well that I recently went to an Irwin model offering nine tools in one — screwdrivers and nut drivers,

as well. It's a hit!

Another indispensable tool for tower work is a cutting tool, a knife. Most everyone chooses the standard box cutter version. But make sure you use one with a retractable blade, and keep spare blades stored in the handle. I've had great success with the Irwin model and their Titanium blades, by the way.

I'd be remiss in talking about old, rusty hardware if I did not mention the only way to deal with it — a product called PB Blaster (a chemical catalyst) shown in **Figure A-9**. This stuff works better to loosen long-rusted hardware than anything I've ever tried. While it was scarce for a while, it's now seemingly available everywhere, including Wal-Mart.

To haul assorted stuff up the tower (what's not carried in one of my belt-side pouches), I use the old faithful Klein Lineman's canvas buckets. I have the oval version (#5144, useful for hand tools — **Figure A-10**), and the more typical, straight-sided bucket (#5103, useful for rotators, and larger items — **Figure A-11**). Lots of folks suggest or mention they use five gallon buckets. But, they're not that practical, in my humble opinion. Their rigid nature, and that pesky metal bail handle, never seem to function well aloft. For storing or carrying items to a job location, sure, but they don't work well in the air.

White lithium grease has many uses. Carry a small tube (**Figure A-12**) up the tower to grease tower legs during assembly. That may make the sections easier to take apart if you need to disassemble the tower.

As described in Chapter 8, a strain gauge or scale used to weigh fish (**Figure A-13**) makes it easy to balance antenna booms. Just suspend the antenna at the boom-to-mast plate, attach the gauge to the light end and pull down until the boom is level. The gauge will tell you how much weight to secure inside that end of the boom.

Figure A-10 — The oval Klein bucket is my choice for tools.

Figure A-11 — The straight-sided Klein bucket is handy for hauling rotators and larger items up the tower.

Figure A-12 — A small tube of white lithium grease is handy for lubricating tower sections during assembly.

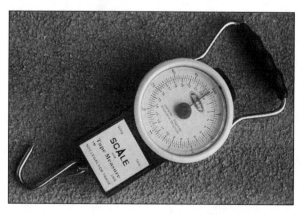

Figure A-13 — The strain gauge is a handy tool for determining how much weight is needed to balance a boom. You can find this handy tool at sporting goods stores.

APPENDIX B

Fasteners

There are bolts, and there are screws. Bolts are held in place by nuts; screws are threaded into one of the pieces being fastened together. The bolt is held stationary, and tightened by turning down the nut. Screws, of course, are tightened by turning the screw head itself. In other words, the application determines the correct terminology.

Plain Talk About Threads

The helical screw thread is amazingly versatile — several different types are found in widespread use. The *Acme* thread, for instance, is a square form thread used on lathes and milling machines to provide straight, directional motion to various sliding rods and cutting heads. A *buttress* thread is designed to withstand severe stresses parallel to the axis of the thread itself. You can find this thread on hose clamps, for instance. Self-tapping sheet metal screws and wood screws come in a variety of threads — without any sort of standardization. All are specialized to suit different materials as well as assembly methods.

One of the first things you notice about bolt threads is that they come in two basic types: *fine pitch* and *coarse pitch*. Each has its place in the world of fasteners. But there is no measurable difference in fatigue resistance

between coarse and fine threads. (As usual, a World War was required to achieve something akin to standardization among nations regarding screw threads, along with calibers of weapons. Interesting history, if you like that sort of thing. By 1948, the military-derived unified fine and unified coarse threaded series, which included a thread angle of 60°, had become standard.)

Bolt Functions

After a few words on threads, let's examine the function of the bolt. Let's start with what a bolt is designed to do. First and foremost, bolts are meant to hold parts together. That's all. They are not designed to serve as pivots, rotational axles, fulcrums or anything else. They should not be used to hold parts in place, either. In other words, bolts should not be used to prevent clamped-together pieces from sliding or moving on or against one another. (For the record, this is what dowels, pilot pins or keys are intended to do.) Bolts are meant to be used as clamps. And only clamps.

Clamps, of course, must remain tight, under all sorts of loads, vibration and stresses. A loose bolt's obviously a poor clamp. A loose bolt will fail, and fail quickly. You may ask: "Okay, what keeps a bolted-together joint tight?" First, here are some things that do not

keep bolted-together joints tight: lock washers, thread locking compounds, safety wires, cotter pins or even elastic stop nuts. This is not meant to demean those parts or products. Or even to condemn modern engineering practices. Which might lead you to infer the mechanical world is about to collapse — if you take my statement literally.

What I mean is that in the workaday world, it's simply not practical to clean and torque every single bolt to precise standards. Often, it's not necessary, either. But if bolts are meant to clamp things together, then the bolt will be internally stressed in tension. Tightening a bolt to a specific torque value actually stretches it and loads it in tension to a set level of stress. Thus, a properly torqued bolt will have the most resistance to a given load for the greatest amount of fatigue cycles. Under-stressed bolts will loosen under load and fail. Over-tightened bolts will fail during installation or prematurely under stress.

What's This Mean, in Practical Terms?

As you tighten the nut onto the bolt, the bolt itself stretches. Male threads elongate; female threads compress. This creates an interference condition, which resists loosening. (Plating, the lack of lubrication, length of the threaded area, all these factor in and contribute to making this seemingly simple task, in a word, difficult.) Normally, this won't affect your tower or antenna project. But, if you're an engine builder, you better believe you'll be measuring things like connecting rod bolts for stretch after installation.

Identification

Hams often walk into their friendly local hardware or home supply store, buy some fasteners to solve their installation problem, and happily go on hamming. Sometimes, they're disappointed — either with the strength of the materials, the plating (or lack of), the wear or something else. Mostly, this is simply a case of "bolt education," which we're trying to overcome here.

Check out **Table B-1,** which gives information about bolts used in industry. The Society of Automotive Engineers (SAE) standard J429, which defines bolt strength, identifies various *grades.* The SAE grade of a bolt is marked on its head, using short radial lines. For Grade 5 and up, the number of lines is two less than the SAE grade (three lines indicate Grade 5, six lines indicate Grade 8). No marking is used for Grade 4 and lower. These grades are typically used for bolts ranging from ¼ inch up 1½ inches in diameter.

Whenever possible, you shouldn't be using anything less than Grade 5 bolts on towers, beams and rotators. That hardware store stuff isn't heat-treated, and typically the plating will rust through in a few months of outdoor exposure. Stainless steel fasteners (more on stainless later) are often

recommended for many applications.

Enterprising hams, believing they're "using the very best." will search out Grade 8 bolts. These are very strong, indeed, but their heat-treating makes them brittle. Unless specified by the manufacturer (for base mounting use, for instance), I wouldn't bother with them. What about buying surplus fasteners? It's not worth it. Unless you know what you're doing (this includes having a micrometer in your hand), it's not worth it to risk life and limb to save a few cents, literally. Counterfeit fasteners are big offshore business. If big companies can be taken in (and they have been), consumers can easily be confused in the surplus store.

Stainless Steel

Now, those few words on stainless steel...a ferrous alloy. The American Society for Metals (ASM) defines stainless steel as any alloy containing at least 10% chromium, with or without other elements. Alloy 304 is an *austenitic* stainless alloy (one of the most common of the four types), and the non-magnetic alloy you'll usually find. Alloy 304 has excellent resistance to corrosion. The ASM publishes volumes on metals; it's a vast topic, if you're interested.

What most hams don't know (having run down to the local hardware store for some stainless hardware) is how to install it properly. Stainless will "gall" as you tighten it — meaning the threads will bind (see the next section). Stainless isn't very strong. For critical uses (like a tower itself) stick with Grade 5 hardware.

Some Tips for Working With Nuts and Bolts

Quite often, I spend considerable time assembling antennas, where I work with lots and lots of various types of hardware. The question once arose, "How long should a bolt be?"

I thought that was a pretty good question for a client to ask — right up there with, "How long should my feed line be?" As with the quick answer to the feed line question ("long enough to reach the shack"), a simple answer to the bolt question may not cover everything you need to consider for your particular installation.

Simply put, there's little point to having more than six threads on anything. National Coarse threaded nuts usually have five threads in them; National Fine threaded nuts have about eight. Why? Because the nut is stronger than the bolt, which will break before the nut will strip. It's just that simple.

Some of you may have heard the old shop adage, "You should have two threads exposed above a nut." Why? Simply because the first two threads of a bolt are sometimes poorly formed. They may not engage the nut properly, overloading the nut, weakening it and causing it to strip. If the first two threads are not holding their share of the load, the other threads in the nut will be overstressed, and the nut may strip.

Galling

Every once in a while, there come those inevitable questions or moments that will yield an idea for my *NCJ* column, along with a chuckle or head-

Table B-1
Common SAE Bolt Grades

Marking	SAE Grade	Size (inches)	Tensile Strength (psi)
A	Grade 1	0.25-1.5	60,000
	Grade 2	0.25-0.75	74,000
		0.875-1.5	60,000
	Grade 4	0.25-1.5	115,000
B	Grade 5	0.25-1	120,000
C	Grade 7	0.25-1.5	133,000
D	Grade 8	0.25-1.5	150,000

No Mark (A)   (B)   (C)   ARRL0585   (D)

shake in wonderment.

Galling — a perennial pesky problem with stainless steel hardware — is a case in point. "Why is it called galling?" one of my clients wanted to know. I admitted that while I didn't know the answer to that, I could explain the process, and why and how it happens. It turns out he didn't care about that; he only wanted to know why it was called what it's called. Saying it's also known as "cold welding," a true oxymoron, swung our conversation back to the process, where I was able to offer an explanation.

Galling takes place when stainless steel connections seemingly "fuse" to and with each other — when a bolt and nut jam or freeze together, for instance. You should know that it's not just stainless hardware that exhibits this. Aluminum, titanium and other alloys that naturally form an oxide surface film can also gall. (But when's the last time you used a bolt made from one of those materials?) This film prevents corrosion, one of the major advantages of such fasteners. But as you assemble them, surface pressure builds up between the threads, and the oxides are broken. The oxide shavings build up, and lock together, which further increases heat and friction, which further expands the metals, and *voila*....galling.

The simplest solution is to keep a lubricant at hand during assembly. A drop of it applied to the threads should completely eliminate galling. Yes, a drop! You don't need to literally lather up the threads. I like to use a tube of lubricant gel ("LubeGel") that contains Teflon. It's small and easily carried in my tool pouch. The gel doesn't get all over everything, like spray lubricants, either. Any lubricant will work—even WD-40, in a pinch, if you'll excuse the pun. The special No-Gall stainless steel nuts and bolts are fine, too, although expensive.

Keeping the hardware cool can help, too. That's often difficult in our work, as you might be surprised to realize how hot hardware can get simply by being out in the sun. You can also slow down as you assemble the bolts and nuts.

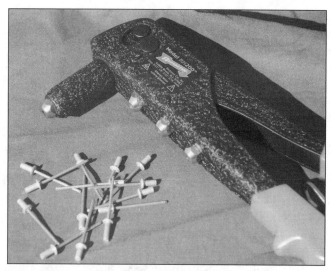

Figure B-1 — A pop rivet tool and aluminum or stainless steel rivets can be useful during antenna assembly

Rivets

Some hams frown on the use of rivets. However, any fastener suitable for aviation and aerospace use (glance out the window the next time you're sitting on some runway, waiting to take off) can be used on a beam, in my opinion.

Pop rivets (the most common variety) are available in various combinations. I recommend the all-aluminum or all-stainless rivet such as those shown in **Figure B-1**. Invest in a quality rivet tool — as you should with all tools. What makes blind or "pop" rivets work so well in fastening element or boom sections together, for instance, is that they expand to fill the hole. Unlike bolts, which must pass through slightly oversized holes, rivets will expand. I recommend *not* buying rivets from the local hardware store. The blister pack-variety are not very strong. I recommend buying Cherry or Avex brand rivets from a fastener supplier.

The holes for rivets should be deburred (admittedly, this is difficult to do inside an element or boom) to ensure proper seating. More importantly, if you want to realize the full strength of a rivet, the maximum rivet diameter should be two to three times the work thickness.

Drills

Of course, all this talk of screws and nuts and rivets, and using them, would not be complete without mentioning something about the holes they go into, or the threads that receive them, to be specific.

To get started, we need a drill (I'm referring to drill bits). Drill sizes are commonly denoted in three ways: smaller drills come in *numbered sizes* from 1 to 60, with the largest being number 1, which is 0.228 inch in diameter, and the smallest, number 60, is 0.040 inch. (That's a lot of drills, Don! Do we really need such a set? And the

Table B-2
Common Numbered Drill Sizes

Number Drill	Diameter (Mils)	Will Clear Screw	Drill for tapping screw hole
10	193.5	10-32	–
18	169.5	8-32	–
21	159	–	10-32
28	140	6-32	–
29	136	–	8-32
35	110	–	6-32
42	93.5	–	4-40

answer is no, probably not. I'd buy several of the most-common sizes, instead. **Table B-2** lists such sizes) You may have seen drills listed with letter sizes, which are commonly called *jobber sizes*, running from A to Z, from 0.234 inch to 0.413 inch. A third system overlaps the numbered and lettered sizes. It's called *fractional sizes*. This is the system you'll see at your local hardware store if you try to go out and buy drill bits. Sizes range from ¹⁄₁₆ to ½ inch.

The bane of drills (like electronics) is heat. Avoid pressure when drilling. Let the bit do the drilling; don't try pushing your way through. Burning the cutting edges of the drill ruins the bit or may cause it to break or bend. Obviously, for best results, any drill should be sharp. Sharpening a drill bit can be difficult, but with practice and the proper tool (several are now available to hold the bit with the proper orientation and angle), you can learn to sharpen bits yourself — just in time to learn something about cutting threads.

Taps and Dies

To cuts threads in material, you use a *tap*. To put threads onto material, you use a *die*. See **Figure B-2**. *Internal* versus *external threads* is the way you remember the distinction. Taps can be purchased three ways — *taper*, *plug* and *bottoming* taps. Taper taps are used to start threads; the tip is ground away for a gradual start. If the material's thick, a plug tap should be used once you've started threading it with a taper tap. Bottoming taps, as their name implies, are used to finish threads to the bottom of a hole that doesn't go all the way through material.

Taps are usually held in a *tap wrench* (various types hold various sizes). But the most important thing to remember is that lubrication is vital to the tap's

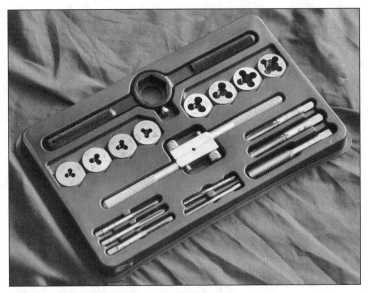

Figure B-2 — A small tap and die set can make the antenna builder's job easier when things need repair.

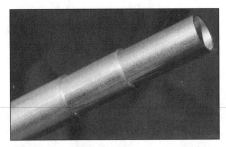

Figure B-3 — Aluminum tubing with 0.058 inch wall thickness telescopes nicely.

life. (You may purchase commercial lubricants, but kerosene works well on aluminum as a lubricant. Always use care and common sense with any machine shop work, but especially when drilling and tapping.) Holes drilled in the material to receive a tap must leave enough material in which to cut the threads. (Table B-2 lists the proper drill sizes used for various common taps.) Move the tap back-and-forth. The backward motion clears the cut metal from the tap. Go slow and don't force the tap.

Have you ever had a dinged-up bolt or cross-

Figure B-4 — Noalox or another antioxidant compound should be used at antenna joints to prevent corrosion.

threaded hole that just wouldn't work? Taps and dies are useful for cleaning up damaged threads as well. It won't always work, but careful application of a tap or die may allow you to save a hard-to-find bolt or machined fitting.

There are, as you might expect, a variety of other fasteners, applications, procedures and methods used in holding things together. Indeed, it's an industry, as a bit of Web searching or a walk through the Yellow Pages will tell you. My remarks are merely an overview, to acquaint you with the basics.

Working with Aluminum Tubing

Materials suitable for building antennas are available from several *QST* advertisers who cater to antenna builders with a variety of tubing, clamps, plates and other hardware. If you live in a large metropolitan area, you may be able to find tubing suppliers in the Yellow Pages.

Supplier catalogs can be an education. In the USA, tubing is available (in 12-foot lengths) in diameters that increase by 0.125 inch (⅛ inch) — ⅜, ½, ⅝ inch outer diameter (OD) and so on. Tubing is available in a variety of wall thicknesses. The secret is to select tubing with a wall thickness of 0.058 inch, which provides perfect telescoping fits (**Figure B-3**). **Table B-3** shows standard tubing sizes. Catalogs will list the various alloys available. Usually, you'll be looking for 6061-T6 or "aircraft aluminum" types of alloys.

Table B-3
Standard Sizes of Aluminum Tubing
6061-T6 (61S-T6) Round Aluminum Tube in 12-foot Lengths

Outer Diameter (inches)	Wall Thickness (inches)	Inside Diameter (inches)	Approx Weight (pounds per foot)	(pounds per length)	Outer Diameter (inches)	Wall Thickness (inches)	Inside Diameter (inches)	Approx Weight (pounds per foot)	(pounds per length)
3/16	0.035	0.117	0.019	0.228	1 1/8	0.035	1.055	0.139	1.668
	0.049	0.089	0.025	0.330		0.058	1.009	0.228	2.736
1/4	0.035	0.180	0.027	0.324	1 1/4	0.035	1.180	0.155	1.860
	0.049	0.152	0.036	0.432		0.049	1.152	0.210	2.520
	0.058	0.134	0.041	0.492		0.058	1.134	0.256	3.072
5/16	0.035	0.242	0.036	0.432		0.065	1.120	0.284	3.408
	0.049	0.214	0.047	0.564		0.083	1.084	0.357	4.284
	0.058	0.196	0.055	0.660	1 3/8	0.035	1.305	0.173	2.076
3/8	0.035	0.305	0.043	0.516		0.058	1.259	0.282	3.384
	0.049	0.277	0.060	0.720	1 1/2	0.035	1.430	0.180	2.160
	0.058	0.259	0.068	0.816		0.049	1.402	0.260	3.120
	0.065	0.245	0.074	0.888		0.058	1.384	0.309	3.708
7/16	0.035	0.367	0.051	0.612		0.065	1.370	0.344	4.128
	0.049	0.339	0.070	0.840		0.083	1.334	0.434	5.208
	0.065	0.307	0.089	1.068		*0.125	1.250	0.630	7.416
1/2	0.028	0.444	0.049	0.588		*0.250	1.000	1.150	14.823
	0.035	0.430	0.059	0.708	1 5/8	0.035	1.555	0.206	2.472
	0.049	0.402	0.082	0.948		0.058	1.509	0.336	4.032
	0.058	0.384	0.095	1.040	1 3/4	0.058	1.634	0.363	4.356
	0.065	0.370	0.107	1.284		0.083	1.584	0.510	6.120
5/8	0.028	0.569	0.061	0.732	1 7/8	0.508	1.759	0.389	4.668
	0.035	0.555	0.075	0.900	2	0.049	1.902	0.350	4.200
	0.049	0.527	0.106	1.272		0.065	1.870	0.450	5.400
	0.058	0.509	0.121	1.452		0.083	1.834	0.590	7.080
	0.065	0.495	0.137	1.644		*0.125	1.750	0.870	9.960
3/4	0.035	0.680	0.091	1.092		*0.250	1.500	1.620	19.920
	0.049	0.652	0.125	1.500	2 1/4	0.049	2.152	0.398	4.776
	0.058	0.634	0.148	1.776		0.065	2.120	0.520	6.240
	0.065	0.620	0.160	1.920		0.083	2.084	0.660	7.920
	0.083	0.584	0.204	2.448	2 1/2	0.065	2.370	0.587	7.044
7/8	0.035	0.805	0.108	1.308		0.083	2.334	0.740	8.880
	0.049	0.777	0.151	1.810		*0.125	2.250	1.100	12.720
	0.058	0.759	0.175	2.100		*0.250	2.000	2.080	25.440
	0.065	0.745	0.199	2.399	3	0.065	2.870	0.710	8.520
1	0.035	0.930	0.123	1.467		*0.125	2.700	1.330	15.600
	0.049	0.902	0.170	2.040		*0.250	2.500	2.540	31.200
	0.058	0.884	0.202	2.424					
	0.065	0.870	0.220	2.640					
	0.083	0.834	0.281	3.372					

*These sizes are extruded; all other sizes are drawn tubes. Shown here are standard sizes of aluminum tubing that are stocked by most aluminum suppliers or distributors in the United States and Canada.

Joining metals such as aluminum can be a joy. Its precision nature and ability to be cut or machined to size is part of that joy. But joining dissimilar metals can be without joy. All metals have an electrolytic potential. Joining them requires you counteract this potential — otherwise you've installed a "battery" (possibly high in the air), which gradually loses material and strength (see the section on Corrosion below). The use of Noalox, Penetrox or other antioxidant compound (**Figure B-4**) is standard procedure when building beams or tower sections. Such lubricants prevent corrosion and will make it easier to get pieces or sections apart later if you have to.

Hose Clamps

Hams who build any type of HF beam antenna will inevitably have to answer the question of how to fasten aluminum tubing pieces together. Joints in the tubing used for booms or elements, which can telescope together quite nicely if the correct tubing sizes are used, must be made solid — both mechanically and electrically.

My first choice is to use stainless steel machine screws at each joint. Some manufacturers and home builders like to cut slots in the larger tubing and then clamp it tightly around the smaller piece that telescopes inside with hose clamps. In this case, I recommend only top-quality worm-gear-type hose clamps with through slots (**Figure B-5**). Clamps with "formed" slots are worthless; they'll strip easily. Stainless steel is the only way to go — for both the clamp and the screw mechanism. (Ideal and Trident are both quality brand names.)

Look for hex-head screws that can be turned either with a screwdriver or (better yet) with a ⁵⁄₁₆-inch nut-driver. I usually try to put the screw itself directly over the slot in the larger tubing before tightening. And, I usually do not try to over-tighten these clamps. Aircraft manuals specify 15 inch-pounds of torque for hose clamps. And that's for a pressurized system. Something less than that will work fine for holding elements together. Obviously, I believe the mechanical joint is the fundamental building block. Electrical continuity can be obtained by driving a stainless steel sheet metal screw through the mechanical connection, if you're truly worried.

Corrosion

Sometimes, when you take apart one of those nifty aluminum antenna joints, you'll discover some fine white powder at the connection. Since it takes a lot of energy to extract aluminum ore from the earth, then create that swell tubing we take for granted, it's easy to see how, over time, this metal will try to return to its natural, corroded state — as it "releases" all the energy we've put into it. In the air, aluminum oxidizes easily, forming aluminum oxide, that white power you encounter.

Another type of corrosion hams encounter frequently is bi-metallic corrosion. This occurs when two metals with the right properties connect, and an electrolyte is present. It's a chemical process (which is sometimes hard for folks to grasp, since we're talking metals, but that's what it is) — just like a battery. Simply put, electrons

Figure B-5 — For a long-lasting installation, look for hose clamps that are all stainless steel.

flow from one metal (called the anode) across the joint to the second metal (called the cathode). Hydrogen gas forms at the surface of the cathodic metal junction. Positive ions left in the anodic metal then oxidize. In bi-metallic joints, the more anodic metal always corrodes away. The electrolyte can be some kind of salt — anything that makes the joint conductive. Acid rain, even dew or salts deposited from your hands, are sufficient to start the process. I've included a chart of galvanic metal activity (**Table B-4**), which shows the electro-potential of these metals or alloys. They are listed by decreasing potential — from most anodic to most cathodic. Generally, it's best to choose those that are close together on the chart for trouble-free results.

For instance, you can see that a junction of aluminum and zinc would be

good, while a joint of aluminum and copper would be bad. Remember that electrical connection means mating surfaces will have microscopic bumps and points where these metals meet. The joint impedance is proportional to the number of such points. Lots of points with little getting in the way means a good joint.

For aluminum, I've found it's best to clean joints as described in the next section. Do not contaminate the clean surface with your fingers after cleaning. Joint pressure is important because oxides start to form immediately after cleaning — the pressure must be great enough to break through this layer. As the metals flex, a phenomenon called "fretting corrosion" occurs, whereby the clean metal part of the open connection oxidizes and builds up. This is why antennas (which worked just fine when you put them up) sometimes seemingly fail all by themselves when up in the air.

Joint compounds such as Noalox or Penetrox can help seal these connections and inhibit electrolytic activity. They are available from electrical supply houses, hardware stores, some home centers, even a couple of antenna manufacturers ship them. These compounds will not last forever, though. They dry out; they harden and crack. Over time and through temper-

Table B-4
Corrosion Between Dissimilar Metals

Anodic — More likely to be attacked	*Cathodic — More Noble, less likely to be attacked*
Magnesium	
Magnesium Alloys	Brass
Zinc	Copper
Aluminum 1100	Bronze
Cadmium	Copper-Nickel Alloys
Aluminum 2024-T4	Stainless Type 430 (Passive)
Steel	Stainless Type 304 (Passive)
Iron	Stainless Type 316 (Passive)
Cast Iron	Silver
Lead-Tin Solders	Graphite
Lead	Gold
Tin	Platinum

ature variations, they will simply flow away from joints. Some sort of finish or overcoat can also be a good idea. Choose something that will flex and resists ultraviolet light.

Cleaning Aluminum

Mention cleaning aluminum and you never fail to rouse some rabble among your audience. *Everyone* has an opinion. Several methods involve steel wool. Alas, this is not a good idea. The wool (even in 0000, very fine grade) will leave behind traces of steel, which will oxidize and rust, creating a far worse problem. There are chemical products that will aid in cleaning.

Never Dull or Gord's Cleaner/Polish are a couple I've used successfully.

The simplest solution is to use Scotch-Brite Industrial Hand Pads (6 × 9 inches). These pads (which are color-coded) are great cleaning tools, and not just for aluminum. I bought a roll of the gray pads five years ago and am not yet a third through it.

Here's the industrial pad breakdown:

- Scotch-Brite Heavy Duty Hand Pad (tan color) — Most durable and aggressive pad for quick removal of dirt and oxidation.
- Scotch Blending Pad (gray color) — Slightly finer abrasive than heavy duty pads. Excellent for

removing scratches on metal, wood or synthetic surfaces.
- Scotch-Brite General Purpose Hand Pad (maroon color) — The most popular Scotch-Brite abrasive because it's coarse enough for cleaning and finishing, but fine enough to produce good surface finishes.
- Scotch-Brite Ultra-Fine Hand Pad (light gray color) — Fine textured for fine finishing of metal, wood, plastics and composites.
- Scotch-Brite Light Duty Hand Pad (white color) — A very mild abrasive for gentle, yet thorough cleaning. Commonly used with liquid detergent.

Random Thoughts and Tips

Tower Standards

You could purchase the EIA/TIA Revision G standard from TIA, but it's ridiculously expensive. It makes more sense to review the *differences* included within the version, compared to earlier ones. There's a PDF file available listing those *differences* online at **www.mei1inc.com/NAB-2003 presentation.pdf**.

In case you are interested, here are the topics covered in Revision G, by chapter:

1) General
2) Loads
3) Analysis
4) Designed strength of structural steel
5) Manufacturing
6) Other structural materials
7) Guy Assemblers
8) Insulators
9) Foundations and anchorages
10) Protective grounding
11) Obstruction and marking
12) Climbing and working facilities
13) Plans, assembly tolerances and marking
14) Maintenance and condition assessment
15) Existing structures

There are 14 annexes that form procurement and user guidelines. These are intended to help engineers in their procurement of antenna supporting structures and antennas designed in accordance with the new standard. Annexes provide amplification and clarification of many of the specifications. The annex listing follows:

A) Procurement and the user guidelines
B) U.S. County listings of design criteria
C) Wind force on typical antennas
D) Twist and sway limitations for microwave antennas
E) Guy rupture
F) Presumptive soil parameters
G) Geotechnical investigations
H) Additional corrosion control
I) Climber attachment anchorages
J) Maintenance and condition assessment
K) Measuring guy tensions
L) Wind speed confessions
M) SI conversion factors
N) References

Some Not-So-Technical Thoughts on Antennas and Hamming in General

Antennas remain one area where hams can both design or create and build something they can consider state-of-the-art. Not too many of us can make something like a modern transceiver in a home workshop, but some simple pieces of wire or aluminum can contribute significantly to the success of that modern transceiver inside your station. All you need is some knowledge, enthusiasm and perseverance. Here's how you can do it.

Knowledge

Books remain the single easiest and most accessible source of antenna information. The most popular source for this favorite topic of ham radio lore is the *ARRL Antenna Book*. It's updated regularly, and while the laws of physics haven't changed, I believe it's a good idea to purchase a copy when a new edition comes out. Today's edition has over 700 pages; it even comes with software.

Another great source for design ideas is *The ARRL Antenna Compendium* series. There are seven volumes available; each contains a wealth of previously unpublished designs. Antennas, transmission lines and propagation are all covered.

Bill Orr's *Radio Handbook* is another must-have title. I also recommend Orr's other antenna books — specifically his works on verticals, quads, beams and wire antennas. Paul Lee's *Vertical Antenna Handbook* is a wealth of information. Jim Lawson's

Yagi Antenna Design, Dave Leeson's *Physical Design of Yagi Antennas,* L. A. Moxon's *HF Antennas For All Locations,* Erwin David's *HF Antenna Collection,* Jacobs & Cohen's *Short-wave Propagation Handbook* and the *Commercial Products* catalog from Rohn are also must-have items in the modern DXer and contester's library. Some of these publications are out of print but often can be found on Internet auction sites, ham radio classified swap sites, flea markets, or through Amazon.com's used book locator service.

The Internet is a virtual library. A surprising number of ham radio links can provide you with fascinating reading, along with a wealth of information and some surprising misinformation. Always seek out a second opinion, especially for advice received from various reflectors.

Enthusiasm

This ought not be a problem for hams, but I do encounter conversations on 2 meter FM that surprise me. (I sometimes simply switch to scanning the channels while driving around the country, going from job to job.) And while this is a complex issue (why wouldn't anyone be excited about our hobby?), I think the answer is a simple one. Our society and culture insulates us from the "old way" of sharing, which was typically via an Elmer or from the necessary "pooling of resources" that club activity entailed. I think the best way to combat this lethargy is to have dedicated Elmers for each newcomer to our hobby. Meaning local clubs will have to become more involved.

Once you have the enthusiasm, the best way to guarantee its continuing existence is to promote operating. And by that I mean HF operating. There is a lot of interesting and exciting stuff to do in ham radio beyond talking with the usual crowd on the local 2 meter repeater. And what gets one's energy and enthusiasm up more than DXing or experiencing an unusual band opening or trying a new mode? Contesting is a big one for me, but that's another

story. I believe I'm enthusiastic enough for two hams!

Perseverance

Dictionaries tell us that *perseverance* means that one continues, steadfast, even when something is difficult or tedious. Ham radio, without a mentor or Elmer, can seem difficult. Certainly, some of the conversations you might encounter can seem tedious. But I would urge anyone reading these lines to consider Stanhope's idea: "Whatever is worth doing at all, is worth doing well." This line from 1746 certainly holds true more than 250 years later.

Try to be a good ham radio operator — which means learning as much as you can about operating, something we all too often take for granted in our hobby. Have the best signal you can, on whatever band, using whatever mode, whenever you're on the air. And make sure you do that — get on the air. You'll find that these three areas inter-relate and drive each other. As your knowledge grows, so will your enthusiasm. If you persevere, you'll find your knowledge increasing, which will create more enthusiasm, and you'll advance through the ranks. See how it works? I hope so!

Wasps ... the Flying Kind

If you spend any time on a tower, you'll eventually encounter these critters, especially during the fall months, when the heat-absorbing steel seems to attract them in great numbers. It's always disconcerting to be climbing, glance upward, and see a swirling swarm of them surrounding the space where you'll soon find yourself.

Like many of us, you'll probably keep climbing, hoping they don't bother you. Most of the time, this will be true. But I have been stung, and probably will be again. The question of what attracts them is unknown, and no so-called wasp expert has stepped forward or provided me with an answer, despite numerous queries. We all

simply assume it's the warming effect from the steel tower that seems to attract them.

Rumors abound about the effects of their stings, however. Let's try to correct these mistakes, some of which can actually be life-threatening. You often hear that taking an antihistamine (Benadryl is a typical example) will stop an allergic reaction. But that reaction (anaphylaxis) is very rapid, and you'd be dead well before the tablet dissolved in your system. If you're allergic to bee stings, you probably already know this.

If you are allergic and must climb, you probably have seen your doctor and have a prescription for an epinephrine self-injector. Of course, you carry that injector with you at all times. You'll have only minutes to react, hopefully allowing yourself time to climb down and get to an ER. The epinephrine will wear off, and the anaphylactic shock may reoccur, but once in the ER, you can receive proper treatment.

Indeed, if you've had an allergic reaction to insect bites or stings, I'd avoid doing tower work during the fall when wasps are around.

Station Design

An Engineered Approach For Today's DX/Contesting Amateur

Paul Rockwell, W3AFM, wrote a series of articles that influenced the way hams thought, not only about DXing in general, but specifically about designing their stations to realize that goal — successfully working DX. The four part series, "Station Design for DX," was published in *QST,* September through December 1966, and remains useful and enjoyable reading today.

What W3AFM set forth was a simplified, yet structured, approach to station design. I believe he borrowed some techniques and terminology from his professional life. Jim Ahlgren, W4RX, who was a young engineering graduate at the time and worked with Paul, agrees with this assessment. Regardless, continuing to follow and use some of his principles, and expanding or enlarging others, we can readily meet some of today's modern station-building needs.

Design

The key word in W3AFM's approach is *design*, something most hams overlook or simply don't consider. Instead, they buy gear (individual pieces and components) and set up stations, often seemingly at random, and *then* decide what they want to do with it. Such an approach is understandable; indeed, it's probably inevitable as your focus within the hobby and your needs develop and change.

For DXing, W3AFM's approach ranked the focus on design as follows:

1) Antenna topics
2) Economics
3) Station configuration (and receiver topics)
4) Propagation quirks and operating tips

That classic approach is just as viable, and just as valuable, to today's ham.

In business, we usually consider Systems Engineering to be an interdisciplinary approach — a means wherein we realize some success. The focus is on defining "customer needs." Functionality, documentation, along with design synthesis and system validation, all while considering the complete problem, will become key elements. For industry, this process usually includes:

■ Operations
■ Performance
■ Testing
■ Manufacturing
■ Cost and Schedule
■ Training and Support

Obviously, such an approach integrates disciplines and specialty groups into a team effort. Development can proceed from concept to production to operation.

As hams, we don't need (usually) to consider the manufacturing sequence (save for a one-time effort). Cost, schedule, training and support are important, but we have more flexibility in these areas than might be found in commercial product development. Yet the ideas behind the process — the critical thinking and analysis — can be not only time-saving, but helpful in reducing costs and guaranteeing performance. However, we should remember that this approach would have both PRO and CON sides:

PRO

■ Those with an engineering background (some hams) are usually very comfortable with the systems approach.
■ The approach is very successful where it's possible to specify the requirements to a high degree.
■ The systems engineering approach requires thorough documentation (after all, it was developed by NATO's command-and-control structure and NASA).

CON

■ The process requires heavy documentation.
■ The process assumes it's possible to arrive at near-perfect documentation that's complete. It further assumes such documentation truly represents the ultimate "requirements."
■ The process assumes the final design can be frozen at some point in its development.

So, let's consider some ways we might utilize this approach in designing a station to be successful today — with an emphasis on antennas for DXing and contest work.

Outside the Shack

Tower(s), antenna(s), rotators and associated hardware should be your primary consideration — realizing the most measurable ROI (return on investment).

For a DXer/contester, angles of radiation are critical — the primary advantages from stacking, or at least having multiple antennas from which to choose. Or, as the old cliche says, "You can never have too many antennas."

Reliability is another critical factor. As is peace of mind — not worrying about failures!

Expense often seems, at first glance, not to have been a factor for some operators. (But appearances can often be deceiving — there often are budgets and limits at work even at the largest super-stations, regardless of what you may think) Volumes could be written on how to budget for station building. Having one is probably the simplest and best advice possible! It will help you focus your priorities and get the most bang for your buck.

Inside the Shack

Ergonomics — relaxed and comfortable posture and positioning. Simply put, this means no wasted motion during the 4 hours or 48 hours you're seated at your radio! While it's often considered "new" or contemporary thinking, Katashi Nose, KH6IJ, pointed out, way back in 1958, the fallacy of sending *without having* your pen or pencil in your hand!

Consider operating techniques or tactics, just for a moment. Remember, the transceiver and the CPU are probably the two greatest influences or factors that have changed the way we operate today.

■ How many readers, for instance, have heard, let alone taken part, in an AM pileup?
■ Once upon a time, we'd tune "the whole band," after CQing!
■ Once upon a time, the Multi-operator, Multi-transmitter category did not exist!
■ Once upon a time, a quota system was used in ARRL's DX contest. Meaning there were limits to the number of DLs or JAs you could work.

Obviously, the times they have changed (and are still a-changin'), as that popular song told us way back in the 1960s. So, history should also contribute to our thinking. As George Santayana said: "Those who do not learn from mistakes of the past are condemned to repeat them..."

And yet, some things remain the same. We just have some more modern tools to analyze things and understand how we're doing things, how change is occurring and how we can measure or gauge its effects. For instance: antenna and terrain modeling, databases of scores/rates/multipliers, historical records (if we care to access and use them), and far, far more sophisticated or better gear — usually all contained in one box or seamlessly integrated and automated.

Consider some specifics as you operate — especially some considerations for using that computer. Note that while the "box" may be a whiz-bang tool, we're now really talking about some pretty mundane, basic things — a chair, a desk, lighting and so forth. These have been part of the ham station forever, but not always used with that whiz-bang computer!

1) Monitor screen is at eye level (±10°)
2) Elbows rest at the side
3) Elbows are bent between 60 to 90°
4) Wrists are in a neutral (straight) position
5) Wrists rest on a rounded tabletop edge
6) Knees are level with or slightly lower than hips
7) Soft seat edge (behind knees)
8) Feet rest on the floor or on a footrest
9) Low and middle back support is provided
10) Seat width 18 inches
11) Seat depth 15 to 17 inches
12) Chair height
13) Desk height
14) Lighting — no shadows or overly bright spots/areas
15) Audio — neither too low nor too loud, so the level is always constant (prefer headphone use with attached microphone)

By now, you should have concluded that *design* is, indeed, the critical word in our entire approach. Ask Google "what is design?" and you'll only get something like 26,700 answers or definitions. Since our focus is high performance radio, our needs will limit some of them — further sharpening our focus.

For instance, we will only have a finite amount of space to use, both outside and inside the shack. Economics will further define and limit our selection of resources. Our "personal" choices (things we favor, such as phone versus CW versus digital modes) will drive our thinking and decisions. Variables outside our control (propagation) will also motivate our thinking and choices.

But then, *voila*, some of those definitions uncovered by Google start to make sense:

- the act of working out the form of something...
- an arrangement, a scheme...
- a blueprint...
- a decorative or artistic work...
- an invention...
- a plan...

So, ultimately, we realize that design is concerned with how things *ought to be* in order to attain goals and to achieve some higher function. In constructive design, a specific artifact is designed. In argumentative design, the design space and the design trade-offs of a class of designs are discussed. In designing our stations — for DX and/or contesting success — we must share and use processes and factors from both constructive and argumentative design in a true, structured-engineering approach.

Efficiency should drive our thinking when building our stations, from inside the shack, to the tower(s) and antenna(s) up in the air. Economics (the practical realities of what we can afford) always further focus those thoughts. It's an approach pioneered by Paul Rockwell, W3AFM, over 40 years ago!

Reference and Inspiration

The virtual library aspects of the World Wide Web remain unsurpassed, allowing you to visit and see what your fellow hams have done to solve problems in their own station-building efforts.

Over the years, I've searched for ideas and solutions (especially good, clear pictures) for various clients and their jobs, and found some of the best ones on the following personal pages.

- **www.ab7e.com** — David Gilbert, AB7E, of Hereford, Arizona (right on the Arizona/Mexico border), has an excellent write-up of his AN Wireless HD-70 tower installation, along with some truly terrific pictures detailing the project, from start to finish.
- **roode.com/k6nr/tower/** — Dana Roode, K6NR, of Irvine, California, has a website with excellent pictures documenting the tower erection at his remote station in the desert. Of special note (truly worth studying) are his "tower permit building notes," which can be found on a separate link at the bottom of the homepage.
- **www.comdac.com** — Duane Durflinger, KX8D, of Hillman, Michigan, has a Web site with *lots* of excellent pictures detailing his crank-up tower installation in the lower Upper Peninsula of Michigan. Of special note are the methods he used in overcoming the very sandy soil conditions encountered.
- **www.n0hr.com** — Pat Rundall, N0HR, of Ames, Iowa, runs a Web site dedicated to ham radio resources, including a vast variety of links and so forth. His own personal tower project pages contain a wealth of information, from start to finish, including some excellent pictures of his

AN Wireless installation. Pat's project is typical, in that he relied on local helpers, ham buddies, family and neighbors to make things happen.

- **k7nv.com** — Kurt Andress, K7NV, of Minden, Nevada, has a very interesting and detailed Web page. Of special interest are his studies of guy wires and guyed towers, all found within "The K7NV Notebook" pages. Other links provide valuable information on Kurt's rebuilding of prop pitch motors, as well.
- **www.k7vc.com** — Dick Flanagan, K7VC, out in western Nevada, has some excellent pictures of his US Tower 72-foot crank up installation, done using a man-lift, along with lots of ham friends helping him stack a variety of beams.
- **www.k4ja.com** — And although it's since been dismantled (I spent a week taking down the antennas and towers, a heart-breaking job), the construction photographs of Paul Hellenberg, K4JA's station in Callao, Virginia, are all still up and very worthwhile viewing. They are truly awe-inspiring and inspirational station-building shots, along with some contesting information!
- **users.erols.com/n3rr/** — Bill Hider, N3RR, of Rockville, Maryland, took a systems engineering approach to building his tower and antenna setup. As a contester, Bill was focused on safety and reliability. Several of his solutions are worth a look.
- **www.k3lr.com**. Tim Duffy, K3LR, in West Middlesex, Pennsylvania (almost in Ohio), has a large multi-multi oriented contest station. There are lots of scores, lots of pictures, lots of contest stuff. But again, don't let all that enthusiasm for competition scare you away if you're not interested in contesting. What's really valuable and useful is examining how Tim has *designed*, and then followed through in *building* his station. I guarantee a visit here will provide you with some ideas!
- **www.anwireless.com** — Web site of AN Wireless Company, of Somerset, Pennsylvania, complete with full technical specifications and descriptions of their self-supporting towers, which are rugged and reliable designs. Perhaps of most interest and use to hams will be the vast arrays of photographs showing (in detail) a wide variety of installations around the world. Procedures and solutions can easily be gleaned from viewing how other folks did what you may wish to do!
- **www.nr5m.com** — George DeMontrond, NR5M's station in Hempstead, Texas, is well on the way to becoming one of the premiere multi-op stations in the USA. John Crovelli, W2GD, and I have been working there together since fall of 2007, rebuilding George's station. Again, the focus is on contesting, but the level of seriousness, enthusiasm and design particulars are worth a look.
- **www.k4za.com** — Shameless self-promotion, I suppose, but this Web site, designed as a blog, may provide you with some interesting reading as I work on various jobs around the country.

APPENDIX

Commercial Products — Application Notes

Preformed Line Products Big Grip Dead Ends are the product of choice for today's tower worker, whether you're using EHS (steel) or a non-conductive guy line (such as Phillystran or Polygon Rod). The following documents from the Preformed Line Products catalog are reprinted with permission. For more information, see **www.preformed.com**.

GUY-GRIP® Dead-end

GUY-GRIP Dead-ends, installed at the top, the breaker and the anchor, provide today's most effective method for securing guy strand. This unique, one-piece dead-end is neat in appearance and free from bolts or high-stress holding devices. The GUY-GRIP Dead-end was the first to offer the cabled loop, a feature that provides more durability, easier tensioning and adaptability to multiple guying.

GUY-GRIP Dead-ends are made of the same material as the strand to which they are applied. They should be used on hardware that is held in a fixed position. The fitting should not be allowed to rotate or spin about the axis of the strand. They should not be used as tools including come-alongs, pulling-in grips, etc.

NOMENCLATURE

Cross-over Marks:

(A)—Indicates starting point for application on smaller diameter fittings.

(B)—Indicates alternate starting point for application on larger diameter fittings.

Cabled Loop: Furnished as standard, all sizes.
Pitch Length: One complete wrap.

Short Leg-Long Leg: Identifies rods belonging to each leg, after application.

Color Code and Length: Assists in identification of strand size, corresponding to tabular information appearing on price page.

Identification Tape: Shows catalog number, nominal sizes.

Nomenclature

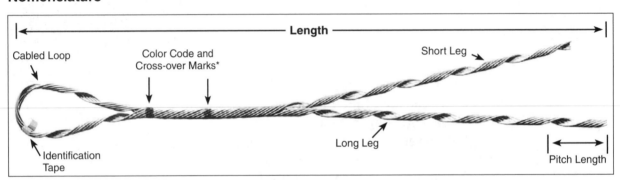

Suggested Hardware Dimensions for Cabled-loop Guy-grip Dead Ends

Nominal Strand Sizes			Seat Dimensions (fig 1 & 2)				
Dead-end Diameter Range (mm)	Galvlvanized/ Galfan Steel	Aluminum-Clad Steel	Max. seat diameter with dead-end at second crossover mark (mm)	Minimum Groove Diameter (mm) (fig. 2)	Minimum Groove Diameter (mm) (fig. 3)	Minimum Hole Diameter* (mm) (fig. 3)	
.123"-.143" (3.1-3.6)	1/8"	—	—	3/16" (4.7)	3/16" (4.7)	1/4" (6.3)	
.144"-.173" (3.7-4.4)	5/32"	—	2½" (64)	1/4" (6.3)	1/4" (6.3)	5/16" (7.9)	
.174"-.203" (4.4-5.2)	3/16"	—	2½" (64)	1/4" (6.3)	1/4" (6.3)	3/8" (9.5)	
.204"-.230" (5.2-5.8)	7/32"	3 #10, 4M3	2½" (64)	5/16" (7.9)	5/16" (7.9)	3/8" (9.5)	
.231"-.259" (5.9-6.6)	1/4"	7 #12, 6M	2½" (64)	5/16" (7.9)	5/16" (7.9)	7/16" (11.0)	
.260"-.291" (6.6-7.4)	9/32"	7 #11, 8M	2½" (64)	3/8" (9.5)	3/8" (9.5)	1/2" (12.7)	
.292"-.336" (7.4-8.5)	5/16"	7 #10, 10 M	2½" (64)	3/8" (9.5)	3/8" (9.5)	9/16" (14.2)	
.337"-.394" (8.6-10)	3/8"	7 #8, 14M, 16M	2½" (64)	7/16" (11.0)	7/16" (11.0)	5/8" (15.8)	
.395"-.474" (10-12)	7/16"	7 #7, 18M, 20M	2½" (64)	1/2" (12.7)	1/2" (12.7)	11/16" (17.4)	
.475"-.515" (12.1-13.1)	**	7 #6	—	9/16" (14.2)	9/16" (14.2)	3/4" (19.0)	
.516"-.570" (13.1-14.5)	**	7 #5, 25M	—	5/8" (15.8)	5/8" (15.8)	15/16" (23.7)	

Figure 1

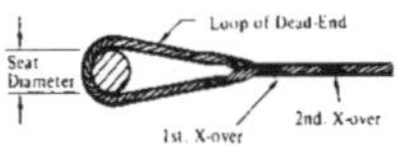

Figure 2

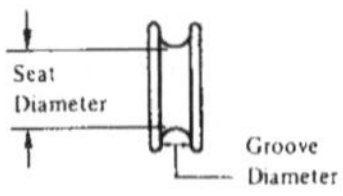

Figure 3

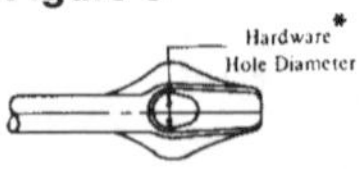

*Depending on geometric shape of the hole, a hole diameter less than specified may be acceptable.
**Use Big-Grip Dead-ends (Refer to Page 20-6).
Note: Alumoweld® is a registered trademark of the Copperweld Company.

GUY-GRIP® Dead-end

Acceptable Fittings

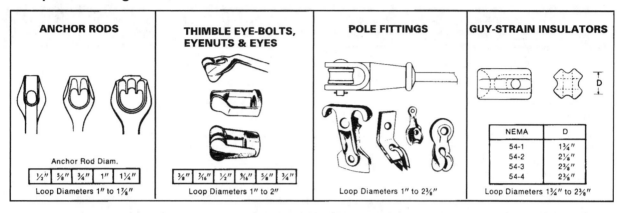

ANCHOR RODS	THIMBLE EYE-BOLTS, EYENUTS & EYES	POLE FITTINGS	GUY-STRAIN INSULATORS

Anchor Rod Diam. ½″ ⅝″ ¾″ 1″ 1¼″
Loop Diameters 1″ to 1⅞″

⅜″ ⁷⁄₁₆″ ½″ ⁹⁄₁₆″ ⅝″ ¾″
Loop Diameters 1″ to 2″

Loop Diameters 1″ to 2⅜″

NEMA	D
54-1	1¾″
54-2	2⅛″
54-3	2⅜″
54-4	2⅜″

Loop Diameters 1¾″ to 2⅜″

For Use on Galvanized Steel Strand

Catalog Number		Size (mm)	Strand Construction	Mean Diameter (mm)	Length (mm)	Color Code	Units Per Carton	Wt./Lbs. Per Carton
B-Coat	C-Coat							
GDE-1102	GDE-2102	³⁄₁₆″ (4.7)	7W	.186″ (4.7)	20″ (508)	Red	100	30
			7W	.195″ (4.7)				
GDE-1104	GDE-2104	¼″ (6.3)	3W	.259″ (6.5)	25″ (635)	Yellow	50	24
			7W	.240″ (6.0)				
GDE-1106	GDE-2106	⁵⁄₁₆″ (7.9)	3W	.312″ (7.9)	31″ (788)	Black	50	39
			7W	.312″ (7.9)				
			7W	.327″ (7.9)				
GDE-1107	GDE-2107	⅜″ (9.5)	3W	.356″ (7.9)	35″ (788)	Orange	50	51
			7W	.360″ (9.1)				
GDE-1108	GDE-2108	⁷⁄₁₆″ (11.0)	7W	.435″ (11.0)	38″ (965)	Green	25	40

Bezinal® Strand*

Catalog Number	Size (mm)	Strand Construction	Mean Diameter (mm)	Length (mm)	Color Code	Units Per Carton	Wt./Lbs. Per Carton
BDE-9102	³⁄₁₆″ (4.7)	7W	.186″ (5)	20″ (508)	Red	100	30
		7W	.195″ (5)				
BDE-9104	¼″ (6.3)	3W	.259″ (7)	25″ (635)	Yellow	50	24
		7W	.240″ (6)				
BDE-9106	⁵⁄₁₆″ (7.9)	3W	.312″ (8)	31″ (788)	Black	50	39
		7W	.312″ (8)				
		7W	.327″ (8)				
BDE-9107	⅜″ (9.5)	3W	.356″ (8)	35″ (788)	Orange	50	51
		7W	.360″ (9)				
BDE-9108	⁷⁄₁₆″ (11.0)	7W	.435″ (11)	38″ (965)	Green	25	40

Notes:
1. Left hand lay standard.
2. Big-Grip Dead-end recommended for guying metal towers and antennas.
* Dead-ends manufactured from Bezinal® Material. Bezinal is a registered trademark of the Bekaert Company.

GUY-GRIP® Dead-end

Aluminum-Clad Steel						
Catalog Number	Mean Diameter (mm)	Nominal Strand Size	Length (mm)	Color Code	Units Per Carton	Wt./Lbs. Per Carton
AWDE-4108	.220" (5.5) .220" (5.5)	4M 3 #10	21" (533)	Green	50	20
AWDE-4110	.242" (6.1) .247" (6.2)	6M 3 #9	24" (610)	yellow	50	20
AWDE-4113	.272" (6.9) .277" (7.0)	8M 3 #8	24" (610)	Blue	50	22
AWDE-4116	.306" (7.7) .306" (7.7) .311" (7.8)	$5/16$"—7 #10 10M 3 #7	26" (660)	Black	50	29
AWDE-4119	.343" (8.7) .343" (8.7) .349" (8.8)	$11/32$"—7 #10 12.5M 3 #6	29" (737)	Yellow	50	41
AWDE-4120	.363" (9.2)	14M	31" (787)	Blue	50	53
AWDE-4122	.385" (9.7) .386" (9.8) .392" (9.9)	$3/8$"—7 #8 16M 3 #5	32" (812)	Orange	50	55
AWDE-4124	.417" (10.5)	18M	34" (864)	Black	25	37
AWDE-4125	.433" (10.9)	$7/16$"—7 #7	36" (914)	Green	25	40
AWDE-4126	.444" (11.2)	20M	37" (940)	Yellow	10	22

Note: Left hand lay standard. Nominal strand size indicates one of various conductors within each range. Big-Grip Dead-end is recommended for guying metal towers and antennas.

Catalog Number	Size (mm)	Strand Construction	Mean Diameter (mm)	Length (mm)	Color Code	Units Per Carton	Wt./Lbs. Per Carton
Stainless Steel for Type 302 Strand							
SDE-5101	$7/32$" (5.5)	3W 7W	.224" (5.6) .216" (5.4)	22" (559)	Blue	100	30
SDE-5102	¼" (6.3)	3W 7W	.259" (6.5) .249" (6.3)	26" (660)	Yellow	50	25
SDE-5103	$9/32$" (7.1)	7W	.279" (7.0)	27" (686)	Black	50	26
SDE-5104	$15/16$" (23.7)	3W 7W	.312" (7.9) .312" (7.9)	31" (787)	Orange	50	41
SDE-5105	$3/8$" (9.5)	3W 7W	.356" (9.0) .360" (9.1)	37" (940)	Green	50	66
Stainless Steel for Type 430 Strand							
SDE-5106	$7/16$" (11.0)	7W	.435" (11.0)	43" (1092)	Red	25	53
Stainless Steel for Type 316 Strand							
SDE-6101	$7/32$" (5.5)	3W 7W	.224" (5.6) .216" (5.4)	22" (559)	Blue	100	29
SDE-6102	¼" (6.3)	3W 7W	.259" (6.5) .249" (6.3)	26" (660)	Yellow	50	25
SDE-6103	$9/32$" (7.1)	7W	.279" (7.0)	27" (686)	Black	50	26
SDE-6504*	$8/16$" (7.9)	3W 7W	.312" (7.9) .312" (7.9)	31" (787)	Orange	50	41
SDE-6105	$3/8$" (9.5)	3W 7W	.356" (9.0) .360" (9.1)	37" (940)	Green	50	66

Note: Left hand lay standard. GUY-GRIP Dead-ends for copper-covered steel are also available.
*These Dead-ends utilize the open helix loop design.

Big-Grip Dead-end

Big-Grip Dead-ends are designed for use with antenna, communications tower, microwave and various guyed structures that require use of large guy strand. They are effective at both the structure top and the anchor bottom.

Big-Grip Dead-ends are left hand lay standard and are applied to the same basic materials as the strand (galvanized strand, aluminum covered strand, except where noted differently).

The Big-Grip Dead-end is designed to be applied quickly in the field, without tools, and usually by one person.

Concentrated stresses in the anchor area are minimized by the cabled loop. Long length helical gripping distributes other stresses uniformly and evenly.

For more detailed information concerning installation guidelines contact Preformed Line Products.

Galvanized Strand

For use on:
- Extra High Strength
- Siemens Martin
- High Strength
- Utilities Grade

Catalog Number	Size (mm)	Strand Construction	Actual Diameter (mm)	BG Per Carton Units	BG Per Carton Wt/ Lbs	Approx. Length (m)	Color Code	Rated Holding Strength	% of Strand's Rated Breaking Strength
BG-2115	½" (12.7)	7W or 19W	.495"(12.5) or .500"(12.7)	20	63	49" (1.24)	Blue	26,900#	(100%)
BG-2116	⁹⁄₁₆" (14.2)	7W or 19W	.564"(14.3) or .565" (14.3)	10	48	55" (1.39)	Yellow	35,000#	(100%)
BG-2111	⅝" (15.8)	7W or 19W	.621"(15.7) or .625"(15.8)	10	65	64" (1.62)	Black	40,200#	(100%)
BG-2112	¾" (19.0)	19W	.750"(19.0)	5	54	76" (1.93)	Orange	58,300#	(100%)
*BG-MS-7023	⅞" (22.2)	19W	.885"(22.4)	3	47	90" (2.28)	Green	79,700#	(100%)
*BG-MS-7047	1" (25.4)	19W or 37W	1.000"(25.4) or 1.001"(25.4)	3	76	125" (3.17)	Blue	104,5000# 92,430#	(100%) (90%)

Note: Left Hand Lay Standard.
*Manufactured or Alumoweld® material. Alumoweld is a registered trademark of the Copperweld Co.

Strand Diameter (mm)	Nominal Strand (mm)	Seat Dimensions Min. (mm))	Seat Dimensions Max. (mm)	Minimum Groove Diameter (mm)	Minimum Hardware Hole Diam. (mm)	Thimble Size (mm)	Pin Diameters Min. (mm)	Pin Diameters Max. (mm)	Double Extra Strong Weight Pipe Nominal Size (mm)	Double Extra Strong Weight Pipe O.D.	Double Extra Strong Weight Pipe I.D.
.475"-.515" (12-18)	½" (12.7)	1⅜" (34.9)	2⅜" (60.3)	⁹⁄₁₆" (14.2)	¾" (19.0)	⅝" (15.8)	1" (25.4)	1⅝" (41.2)	1¼" (31.7)	1.66	.896
.516"-.570" (13-14)	⁹⁄₁₆" (14.2)	1½" (38.1)	2⅝" (66.6)	⅝" (15.8)	¹⁵⁄₁₆" (23.7)	⅝" (15.8)	1⅛" (28.5)	1⅝" (41.2)	1¼" (31.7)	1.66	.896
.571"-.635" (15-16)	⅝" (15.8)	2" (50.8)	2⅝" (66.6)	¾" (19.0)	1" (25.4)	¾" (19.0)	1½" (38.1)	1⅞" (47.6)	1¼" (31.7)	1.66	.896
.636"-.772" (16-20)	¾" (19.0)	2½" (63.5)	3⅛" (79.3)	⅞" (22.2)	1³⁄₁₆" (30.1)	⅞" (22.0)	1⅞" (47.6)	2⅛" (53.9)	1½" (38.1)	1.9	1.1
.773"-.868" (20-22)	–	2½" (63.5)	3⅝" (92.0)	1" (25.4)	1⅜" (34.9)	1" (25.4)	2" (50.8)	2⅜" (60.3)	2" (50.8)	2.375	1.503
.869"-1.024" (22-26)	1" (25.4)	3" (76.2)	4⅛" (104.7)	1" (25.4)	1⅜" (34.9)	1⅛"-1¼" (28.5-31.7)	2⅜" (60.3)	2¾" (69.8)	2" (50.8)	2.375	1.503
1.025"-1.27" (26-32)	–	3½" (88.9)	5⅛" (130.1)	1⅜" (34.9)	1¾" (44.4)	1¼"-1⅜" (31.7-34.9)"	2¾" (69.8)	3¼" (82.5)	2½" (63.5)	2.875	1.771
1.30" (33)	–	4" (101.6)	5⅛" (130.1)	1⅜" (34.9)	1¹⁵⁄₁₆" (49.1)	1⅜"-1½" (34.9-38.1)	2⅞" (73.0)	3⅜" (85.7)	2½" (63.5)	2.875	1.771

Figure 1.	Figure 2.	Figure 3.	Figure 4.	Figure 5.	Figure 6.

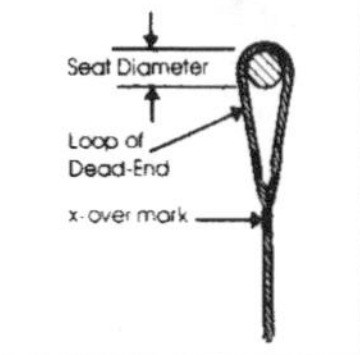

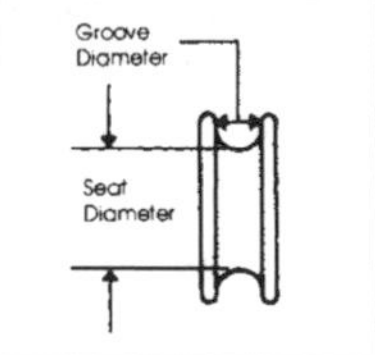

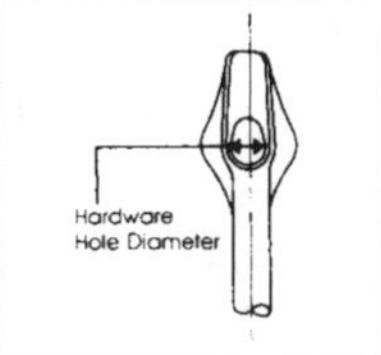

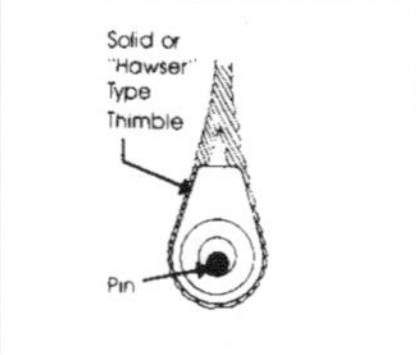

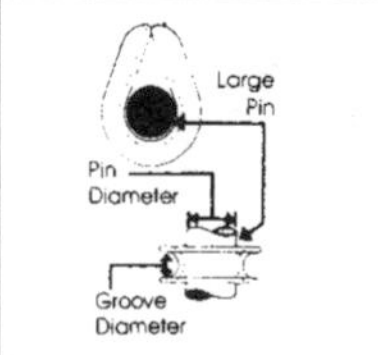

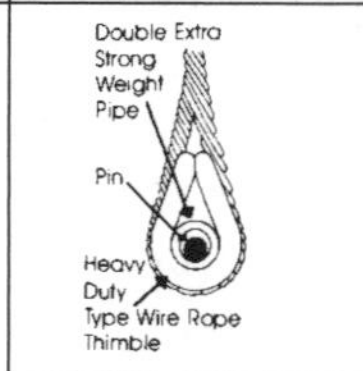

Big-Grip Dead-end

For Use On: Aluminum Covered Steel Strand

Catalog Number	Strand Diameter Range		Nominal Strand Size	BG Per Carton		Approx. Length		Color Code	Rated Holding Strength	% of Strand's Rated Breaking Strength
	Min. Inches	Max. Inches		Units	Wt.-Lbs.	Inches	Meters			
BG-4168	.475	.494	7#6	25	60	42	1.06	Blue	22,730#	(100%)
BG-4169	.495	.515	19#10	25	62	44	1.11	Green	27,190#	(100%)
BG-4170	.516	.536	25M	20	66	47	1.19	Red	25,000#	(100%)
BG-4171	.537	.555	7#5	20	67	48	1.21	Yellow	27,030#	(100%)
BG-4172	.556	.570	—	15	68	49	1.24	Blue	33,330#	–
BG-4173	.571	.591	19#9	20	68	50	1.27	Orange	34,290#	(100%)
BG-4174	.592	.612	—	15	50	50	1.27	Green	34,500#	–
BG-4175	.613	.635	—	10	49	54	1.37	Yellow	45,000#	–
BG-4176	.636	.661	19#8	10	50	56	1.42	Black	43,240#	(100%)
BG-4177	.662	.686	19 x .1363"	10	66	59	1.49	Blue	47,400#	(100%)
BG-4178	.687	.712	—	10	68	61	1.54	Red	54,200#	–
BG-4179	.713	.741	19#7 37#10	10	70	63	1.60	Black	51,730# 50,300#	(100%) (95%)
BG-4180	.742	.772	19 x .1499"	5	41	71	1.80	Yellow	54,300#	–

Catalog Number	Actual Diameter Inches	Nominal Strand Size	BG Per Carton		Approx. Length		Color Code	Rated Holding Strength	% of Strand's Rated Breaking Strength
			Units	Wt.-Lbs.	Inches	Inches			
BG-4181	.792	19 x .1584"	5	50	80	2.03	Blue	59,000#	–
BG-4183	.801, .810, .827	37#9 19#6 19 x .1660"	5	69	84	2.13	Green	63,430# 61,700# 63,000#	(95%) (100%) (100%)
BG-4185	.849, .850, .866	37 x .1213" 19 x .170" 19 x .1732" 37 x .1237"	5	68	87	2.21	Black	71,250# 66,000# 68,500# 74,100#	(95%) (100%) (100%) (95%)
BG-4186	.899	37#8	5	76	91	2.31	Yellow	80,000#	(95%)
BG-4187	.910, .934	19#5 19 x .1868"	5	78	93	2.36	Blue	73,350# 75,000#	(100%) (100%)
BG-4188	.981	37# .1401"	4	55	95	2.41	Red	90,250#	(95%)
BG-4189	1.01	37#7	4	85	106	2.74	Green	90,600#	(90%)
BG-4190	1.10	37 x .1571"	3	87	117	2.97	Black	101,700#	(90%)
BG-4191	1.134	37#6	3	86	120	3.04	Yellow	108,200#	(90%)
BG-4192	1.27	37#5	2	87	151	3.83	Red	127,000#	(89%)

Note: Left Hand Lay Standard.

Phillystran is an excellent choice for non-conductive guy lines for your ham tower. Big Grips are available, and provide an easy way to attach this line. Here is the relevant data sheet covering the products. The following Phillystran Technical Bulletin is reprinted with permission. For more information, see **www.phillystran.com**.

Technical Bulletin

PHILLYSTRAN® HPTG-I HIGH PERFORMANCE BROADCAST TOWER & ANTENNA GUY LINES ARAMID

163-2/06

The excellent dielectric properties of aramid fiber plus its strength-to-weight ratio five times greater than steel plus its low-stretch, low-creep characteristics add up to an ideal material for broadcast tower guys. Today's Phillystran tower guying systems based on aramid fiber are the result of long years of experience. Starting in 1973, Phillystran's tower installations include megawatt arrays as well as towers up to 1300 feet high. Phillystran HPTG-I systems eliminate problems associated with EHS steel cable such as:

- Electromagnetic interference (EMI)
- Radio frequency interference (RFI)
- Signal suppression
- Directional irregularities
- Zapping and white-noise arcing associated with ceramic insulators interference with TV reception near broadcast sites

PART NUMBER	BREAK STRENGTH		DIAMETER		WEIGHT		CORONA SOCKET	REEL LENGTH
	LB	kN	IN	mm	LBS/1000 FT	kg/km	PART NUMBER	FT
HPTG 1200I	1,200	5.3	0.17	4	11	16	CS1200	10,000
HPTG 2100I	2,100	9.3	0.22	6	18	27	CS2100	10,000
HPTG 4000I	4,000	18	0.30	8	33	50	CS4000	10,000
HPTG 6700I	6,700	30	0.37	9	50	75	CS6700	10,000
HPTG 11200I	11,200	50	0.44	11	70	105	CS11200	5,000
HPTG 15400I	15,400	69	0.51	13	95	140	CS15400	5,000
HPTG 20800I	20,800	93	0.57	14	115	170	CS20800	5,000
HPTG 27000I	27,000	120	0.65	17	150	225	CS27000	5,000
HPTG 35000I	35,000	156	0.69	18	170	250	CS35000	4,000
HPTG 42400I	42,400	189	0.84	21	230	350	CS42400	3,000
HPTG 58300I	58,300	259	0.96	24	300	450	CS58300	2,000
HPTG 85000I	85,000	378	1.14	29	420	630	CS85000	2,000
HPTG 130000I	130,000	578	1.56	40	740	1100	CS130000	1,000
HPTG 200000I	200,000	890	1.87	47	1050	1560	CS200000	800
HPTG 252000I	252,000	1,121	2.08	53	1290	1920	CS252000	600

Weights and Dimensions can vary

CAUTION: Break Strength: The breaking strength of a rope is the load at which a new rope will break when tested under laboratory conditions. Break strength should not be mistaken for safe working load. **Safe Working Load:** Because of the wide range of rope use, rope condition and the degree of risk of life or property, it is not possible to make a blanket recommendation for safe working load. It is ultimately dependent on the rope user to determine what percentage of break strength is their own safe working load. **Wear:** Ropes wear out with use; the more severe the usage, the greater the wear. It is often not possible to detect wear on a rope by visible signs alone. Therefore, it is recommended that the rope user determine a retirement criteria for ropes in their application. For assistance in developing safe working load and retirement criteria for each application please call or write Phillystran, Inc.

All printed statements, data and recommendations are based on reliable information and tests, and are presented without any guarantee or warranty. Statements regarding the use of Phillystran, Inc.'s products and processes are not to be constructed as recommendations for use in violation of any applicable laws, regulations or patent rights. © All rights reserved.

Phillystran, Inc.
151 Commerce Drive
Montgomeryville, PA 18936-9628 USA
Phone: 215-368-6611 Fax: 215-362-7956
Web: www.phillystran.com
E-mail: info@phillystran.com

Phillystran Europe BV
Kralingseweg 264B
3066 RA Rotterdam, The Netherlands
Phone: +31 (0) 10 8485924
Web: www.phillystran.nl
E-mail: jkn@phillystran.nl

The Rohn Product Catalog can serve more-or-less as your "Bible," in terms of guyed tower construction. It's well worth the expense of buying a copy, as well as the time and trouble of sitting down and reading through the various sections. You'll learn a lot just by doing so. For more information, see **www.rohnnet.com**.

Here are some relevant illustrations, reprinted with permission, pertaining to some typical ham installations of 25G and 45G towers:

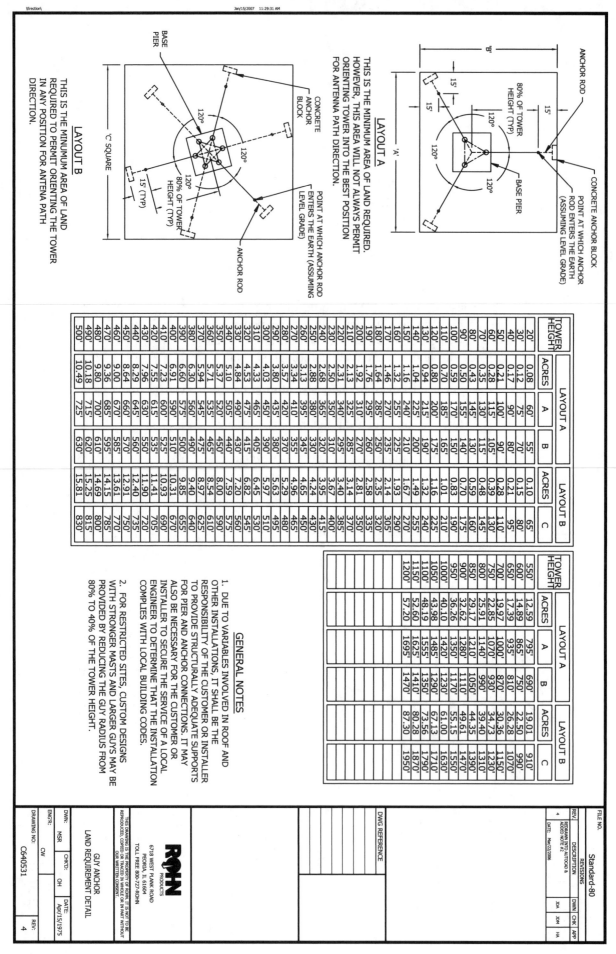

LAYOUT A

THIS IS THE MINIMUM AREA OF LAND REQUIRED. HOWEVER, THIS AREA WILL NOT ALWAYS PERMIT ORIENTING TOWER INTO THE BEST POSITION FOR ANTENNA PATH DIRECTION.

LAYOUT B

THIS IS THE MINIMUM AREA OF LAND REQUIRED TO PERMIT ORIENTING THE TOWER IN ANY POSITION FOR ANTENA PATH DIRECTION.

TOWER HEIGHT	LAYOUT A			LAYOUT B	
	ACRES	A	B	ACRES	C
20'	0.08	60'	55'	0.10	65'
30'	0.12	75'	70'	0.15	80'
40'	0.17	90'	80'	0.21	95'
50'	0.21	100'	90'	0.28	110'
60'	0.28	115'	105'	0.39	130'
70'	0.35	130'	115'	0.48	145'
80'	0.43	145'	130'	0.59	160'
90'	0.50	155'	140'	0.70	175'
100'	0.59	170'	150'	0.83	190'
110'	0.70	185'	165'	1.01	210'
120'	0.80	200'	175'	1.16	225'
130'	0.94	215'	190'	1.32	240'
140'	1.04	225'	200'	1.49	255'
150'	1.16	240'	210'	1.67	270'
160'	1.32	255'	225'	1.93	290'
170'	1.46	270'	235'	2.14	305'
180'	1.64	285'	250'	2.35	320'
190'	1.76	295'	260'	2.58	335'
200'	1.92	310'	270'	2.81	350'
210'	2.13	325'	285'	3.14	370'
220'	2.31	340'	295'	3.40	385'
230'	2.50	350'	310'	3.67	400'
240'	2.68	365'	320'	3.95	415'
250'	2.88	380'	330'	4.24	430'
260'	3.13	395'	345'	4.65	450'
270'	3.34	410'	355'	4.96	465'
280'	3.57	420'	370'	5.29	480'
290'	3.80	435'	380'	5.63	495'
300'	4.03	450'	390'	5.97	510'
310'	4.33	465'	405'	6.45	530'
320'	4.55	475'	415'	6.82	545'
330'	4.84	490'	430'	7.20	560'
340'	5.10	505'	440'	7.59	575'
350'	5.37	520'	450'	8.00	590'
360'	5.71	535'	465'	8.54	610'
370'	5.94	545'	475'	8.97	625'
380'	6.30	560'	490'	9.40	640'
390'	6.60	575'	500'	9.85	655'
400'	6.91	590'	510'	10.31	670'
410'	7.23	600'	525'	10.93	690'
420'	7.55	615'	535'	11.41	705'
430'	7.96	630'	550'	11.90	720'
440'	8.29	645'	560'	12.40	735'
450'	8.64	660'	570'	12.91	750'
460'	9.00	670'	585'	13.61	770'
470'	9.36	685'	595'	14.15	785'
480'	9.80	700'	610'	14.69	800'
490'	10.18	715'	620'	15.25	815'
500'	10.49	725'	630'	15.81	830'

TOWER HEIGHT	LAYOUT A			LAYOUT B	
	ACRES	A	B	ACRES	C
550'	12.59	795'	690'	19.01	910'
600'	14.89	865'	750'	22.50	990'
650'	17.39	935'	810'	26.28	1070'
700'	19.97	1000'	870'	30.36	1150'
750'	22.85	1070'	930'	34.73	1230'
800'	25.91	1140'	990'	39.40	1310'
850'	29.17	1210'	1050'	44.35	1390'
900'	32.62	1280'	1110'	49.61	1470'
950'	36.26	1350'	1170'	55.15	1550'
1000'	40.10	1420'	1230'	61.00	1630'
1050'	43.98	1485'	1290'	67.13	1710'
1100'	48.19	1555'	1350'	73.56	1790'
1150'	52.60	1625'	1410'	80.28	1870'
1200'	57.20	1695'	1470'	87.30	1950'

GENERAL NOTES

1. DUE TO VARIABLES INVOLVED IN ROOF AND OTHER INSTALLATIONS, IT SHALL BE THE RESPONSIBILITY OF THE CUSTOMER OR INSTALLER TO PROVIDE STRUCTURALLY ADEQUATE SUPPORTS FOR PIER AND ANCHOR CONNECTIONS. IT MAY ALSO BE NECESSARY FOR THE CUSTOMER OR INSTALLER TO SECURE THE SERVICE OF A LOCAL ENGINEER TO DETERMINE THAT THE INSTALLATION COMPLIES WITH LOCAL BUILDING CODES.

2. FOR RESTRICTED SITES, CUSTOM DESIGNS WITH STRONGER MASTS AND LARGER GUYS MAY BE PROVIDED BY REDUCING THE GUY RADIUS FROM 80% TO 40% OF THE TOWER HEIGHT.

ROHN PRODUCTS
6718 WEST PLANK ROAD
PEORIA, IL 61604
TOLL FREE 800-727-ROHN

THIS DRAWING IS THE PROPERTY OF ROHN, IT IS NOT TO BE REPRODUCED, COPIED OR TRACED IN WHOLE OR IN PART WITHOUT OUR WRITTEN CONSENT.

GUY ANCHOR
LAND REQUIREMENT DETAIL

DRAWING NO: CG40531
REV: 4

DWN: MSR
CHKD: OH
ENGR: CW

DATE: Apr/15/1975

FILE NO.
Standard-80

REVISIONS
REV.	DESCRIPTION	DATE	DWN	CHK	APP
4	REDRAWN INTO AUTOCAD & ADDED NOTE #2	Mar/22/2006	JDA	JDM	HA

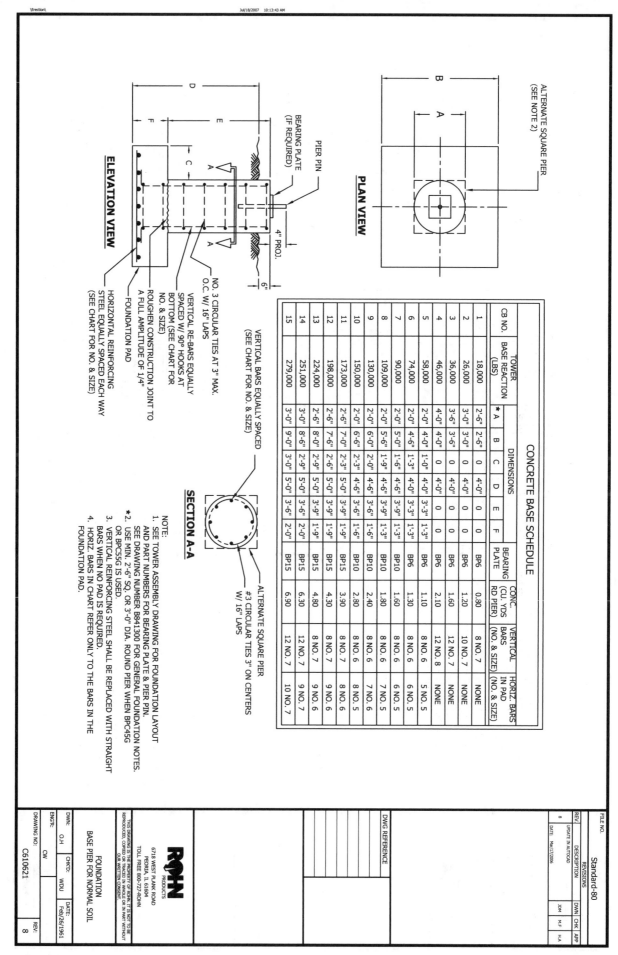

PLAN VIEW

ELEVATION VIEW

SECTION A-A

ALTERNATE SQUARE PIER (SEE NOTE 2)

PIER PIN

BEARING PLATE (IF REQUIRED)

4" PROJ.

6"

ALTERNATE SQUARE PIER #3 CIRCULAR TIES 3" ON CENTERS W/ 16" LAPS

NO. 3 CIRCULAR TIES AT 3" MAX. O.C. W/ 16" LAPS

VERTICAL RE-BARS EQUALLY SPACED W/ 90° HOOKS AT BOTTOM (SEE CHART FOR NO. & SIZE)

ROUGHEN CONSTRUCTION JOINT TO A FULL AMPLITUDE OF 1/4"

FOUNDATION PAD

HORIZONTAL REINFORCING STEEL EQUALLY SPACED EACH WAY (SEE CHART FOR NO. & SIZE)

VERTICAL BARS EQUALLY SPACED (SEE CHART FOR NO. & SIZE)

CONCRETE BASE SCHEDULE

CB NO.	TOWER BASE REACTION (LBS)	*A	B	C	D	E	F	BEARING PLATE RD PIER	CONC. (CU. YDS) RD PIER	VERTICAL BARS (NO. & SIZE)	HORIZ. BARS IN PAD (NO. & SIZE)
1	18,000	2'-6"	2'-6"	0	4'-0"	0	0	BP6	0.80	8 NO. 7	NONE
2	26,000	3'-0"	3'-0"	0	4'-0"	0	0	BP6	1.20	10 NO. 7	NONE
3	36,000	3'-6"	3'-6"	0	4'-0"	0	0	BP6	1.60	12 NO. 7	NONE
4	46,000	4'-0"	4'-0"	0	0	0	0	BP6	2.10	12 NO. 8	NONE
5	58,000	2'-0"	4'-0"	1'-0"	4'-0"	0	1'-3"	BP6	1.10	12 NO. 8	5 NO. 5
6	74,000	2'-0"	4'-6"	1'-3"	4'-0"	0	1'-3"	BP6	1.30	8 NO. 6	6 NO. 5
7	90,000	2'-0"	5'-0"	1'-6"	4'-0"	0	1'-3"	BP10	1.60	8 NO. 6	6 NO. 5
8	109,000	2'-0"	5'-6"	1'-9"	4'-6"	3'-9"	1'-3"	BP10	1.80	8 NO. 6	7 NO. 5
9	130,000	2'-0"	6'-0"	2'-0"	4'-6"	3'-9"	1'-6"	BP10	2.40	8 NO. 6	7 NO. 6
10	150,000	2'-0"	6'-6"	2'-3"	4'-6"	3'-6"	1'-6"	BP10	2.80	8 NO. 6	8 NO. 6
11	173,000	2'-6"	7'-0"	2'-3"	5'-0"	3'-9"	1'-9"	BP15	3.90	8 NO. 7	8 NO. 6
12	198,000	2'-6"	7'-6"	2'-6"	5'-0"	3'-9"	1'-9"	BP15	4.30	8 NO. 7	9 NO. 6
13	224,000	2'-6"	8'-0"	2'-9"	5'-0"	3'-9"	1'-9"	BP15	4.80	8 NO. 7	9 NO. 6
14	251,000	3'-0"	8'-6"	2'-9"	5'-0"	3'-6"	2'-0"	BP15	6.30	12 NO. 7	9 NO. 7
15	279,000	3'-0"	9'-0"	3'-0"	5'-0"	3'-6"	2'-0"	BP15	6.90	12 NO. 7	10 NO. 7

DIMENSIONS

NOTE:
1. SEE TOWER ASSEMBLY DRAWING FOR FOUNDATION LAYOUT AND PART NUMBERS FOR BEARING PLATE & PIER PIN. SEE DRAWING NUMBER B841300 FOR GENERAL FOUNDATION NOTES.
*2. USE MIN. 2'-6" SQ. OR 3'-0" DIA. ROUND PIER WHEN BPC45G OR BPC5SG IS USED.
3. VERTICAL REINFORCING STEEL SHALL BE REPLACED WITH STRAIGHT BARS WHEN NO PAD IS REQUIRED.
4. HORIZ. BARS IN CHART REFER ONLY TO THE BARS IN THE FOUNDATION PAD.

ROHN PRODUCTS
6718 WEST PLANK ROAD
PEORIA, IL 61604
TOLL FREE 800-727-ROHN

FOUNDATION
BASE PIER FOR NORMAL SOIL

DWN: O.H	CHKD: WDU	
ENGR: CW	DATE: Feb/26/1961	

DRAWING NO: C610621 | REV: 8

Standard-80

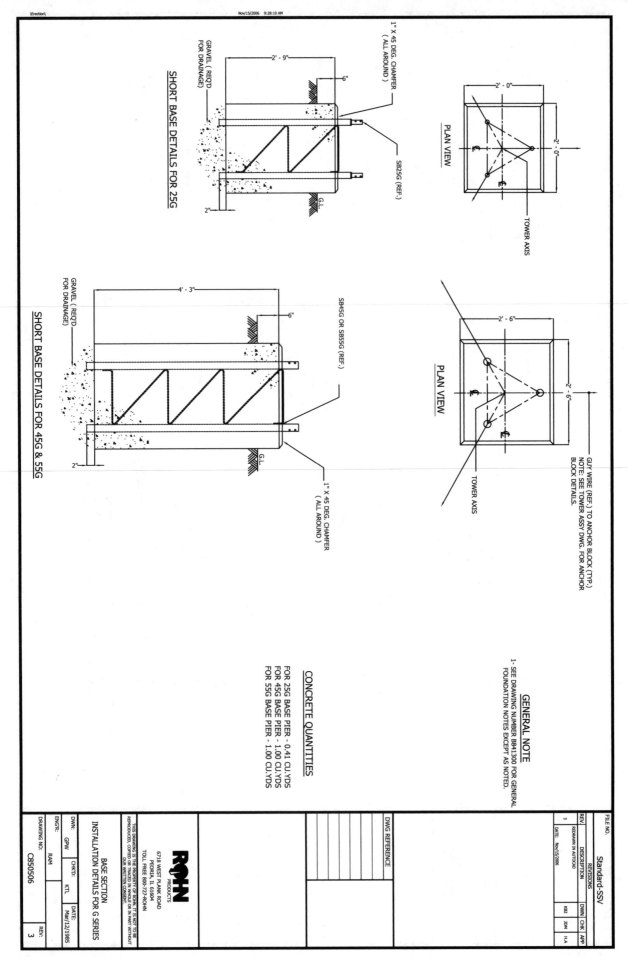

SHORT BASE DETAILS FOR 25G

SHORT BASE DETAILS FOR 45G & 55G

PLAN VIEW

PLAN VIEW

GENERAL NOTE
1- SEE DRAWING NUMBER B841300 FOR GENERAL FOUNDATION NOTES EXCEPT AS NOTED.

CONCRETE QUANTITIES
FOR 25G BASE PIER - 0.41 CU.YDS
FOR 45G BASE PIER - 1.00 CU.YDS
FOR 55G BASE PIER - 1.00 CU.YDS

ROHN
PRODUCTS
6718 WEST PLANK ROAD
PEORIA, IL 61604
TOLL FREE 800-727-ROHN

BASE SECTION
INSTALLATION DETAILS FOR G SERIES

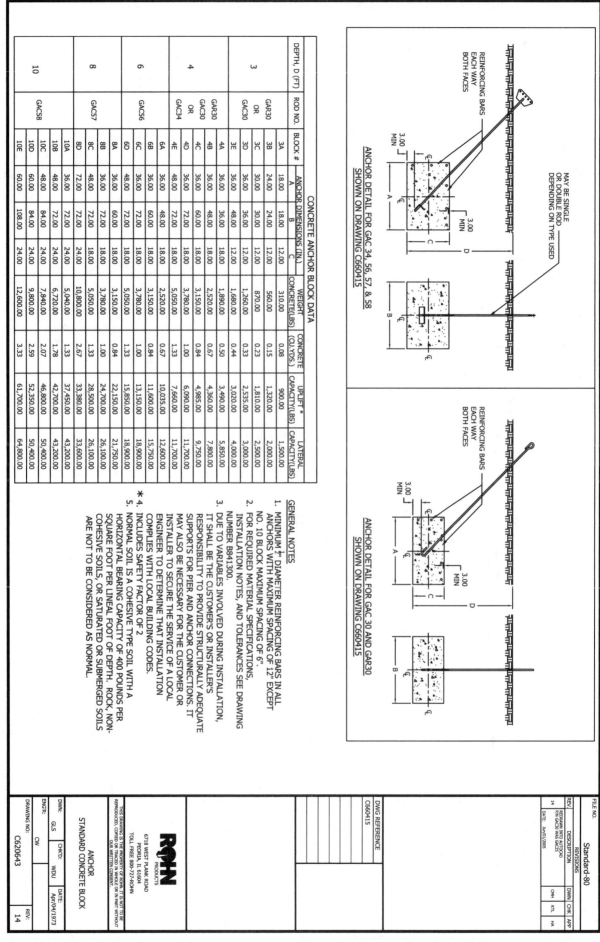

ANCHOR DETAIL FOR GAC 34, 56, 57, & 58 SHOWN ON DRAWING C660415

ANCHOR DETAIL FOR GAC 30 AND GAR30 SHOWN ON DRAWING C660415

MAY BE SINGLE OR DOUBLE ROD DEPENDING ON TYPE USED

REINFORCING BARS EACH WAY BOTH FACES

3.00 MIN

CONCRETE ANCHOR BLOCK DATA

DEPTH, D (FT)	ROD NO.	BLOCK #	ANCHOR DIMENSIONS (IN.) A	B	C	WEIGHT CONCRETE (LBS)	CONCRETE (CU.YDS.)	UPLIFT * CAPACITY (LBS)	LATERAL CAPACITY (LBS)
3	GAR30 OR GAC30	3A	18.00	18.00	12.00	310.00	0.08	900.00	1,500.00
		3B	24.00	24.00	12.00	560.00	0.15	1,320.00	2,500.00
		3C	30.00	30.00	12.00	870.00	0.23	1,810.00	3,000.00
		3D	36.00	36.00	12.00	1,260.00	0.33	2,535.00	4,000.00
		3E	36.00	48.00	12.00	1,680.00	0.44	3,020.00	4,000.00
4	GAR30 OR GAC34	4A	36.00	36.00	18.00	1,890.00	0.50	3,490.00	5,850.00
		4B	36.00	48.00	18.00	2,520.00	0.67	4,360.00	7,800.00
		4C	36.00	60.00	18.00	3,150.00	0.84	4,985.00	9,750.00
		4D	36.00	72.00	18.00	3,780.00	1.00	6,090.00	11,700.00
		4E	48.00	72.00	18.00	5,050.00	1.33	7,660.00	11,700.00
6	GAC56	6A	36.00	48.00	18.00	2,520.00	0.67	10,035.00	12,600.00
		6B	36.00	60.00	18.00	3,150.00	0.84	11,600.00	15,750.00
		6C	36.00	72.00	18.00	3,780.00	1.00	13,150.00	18,900.00
		6D	48.00	72.00	18.00	5,050.00	1.33	15,850.00	18,900.00
8	GAC57	8A	36.00	60.00	18.00	3,150.00	0.84	18,900.00	18,900.00
		8B	36.00	72.00	18.00	3,780.00	1.00	22,150.00	21,750.00
		8C	48.00	72.00	18.00	5,050.00	1.33	24,700.00	26,100.00
		8D	72.00	72.00	24.00	10,800.00	2.67	28,500.00	26,100.00
10	GAC58	10A	36.00	72.00	24.00	5,040.00	1.33	37,450.00	43,200.00
		10B	48.00	72.00	24.00	6,720.00	1.78	42,700.00	43,200.00
		10C	48.00	84.00	24.00	7,840.00	2.07	46,800.00	50,400.00
		10D	60.00	84.00	24.00	9,800.00	2.59	52,350.00	50,400.00
		10E	60.00	108.00	24.00	12,600.00	3.33	61,700.00	64,800.00

GENERAL NOTES

1. MINIMUM $\frac{1}{2}$" DIAMETER REINFORCING BARS IN ALL ANCHORS WITH MAXIMUM SPACING OF 12" EXCEPT NO. 10 BLOCK MAXIMUM SPACING OF 6".
2. FOR REQUIRED MATERIAL SPECIFICATIONS, INSTALLATION NOTES, AND TOLERANCES SEE DRAWING NUMBER B841300.
3. DUE TO VARIABLES INVOLVED DURING INSTALLATION, IT SHALL BE THE CUSTOMER'S OR INSTALLER'S RESPONSIBILITY TO PROVIDE STRUCTURALLY ADEQUATE SUPPORTS FOR PIER AND ANCHOR CONNECTIONS. IT MAY ALSO BE NECESSARY FOR THE CUSTOMER OR INSTALLER TO SECURE THE SERVICE OF A LOCAL ENGINEER TO DETERMINE THAT INSTALLATION COMPLIES WITH LOCAL BUILDING CODES.
* 4. INCLUDES SAFETY FACTOR OF 2
5. NORMAL SOIL IS A COHESIVE TYPE SOIL WITH A HORIZONTAL BEARING CAPACITY OF 400 POUNDS PER SQUARE FOOT PER LINEAL FOOT OF DEPTH. ROCK, NON-COHESIVE SOILS, OR SATURATED OR SUBMERGED SOILS ARE NOT TO BE CONSIDERED AS NORMAL.

FILE NO. Standard-80

REVISIONS

REV	DESCRIPTION	DWN	CHK	APP
14	REDRAWN INTO AUTOCAD P/N GAC76 WAS GAC57 DATE: Jun/01/2005	KTL		HA

DWG REFERENCE C660415

THIS DRAWING IS THE PROPERTY OF ROHN. IT IS NOT TO BE REPRODUCED, COPIED OR TRACED, IN WHOLE OR IN PART WITHOUT OUR WRITTEN CONSENT.

ROHN PRODUCTS
6718 WEST PLANK ROAD
PEORIA, IL 61604
TOLL FREE 800-727-ROHN

ANCHOR
STANDARD CONCRETE BLOCK

DWN: GLS   CHKD: WDU   DATE: Apr/04/1973
ENGR: CW
DRAWING NO: C620643   REV: 14

ROHN ANCHORS

Type EJ equalizer plates are used with eye and jaw turnbuckles.

Type EE equalizer plates are supplied in pairs for eye and eye turnbuckles.

Part number suffixes 3, 5 and 55 denote 3 or 5 holes in plates.

Type GAC30 and GAC305 rods are supplied with type EP25343 or EP25345 equalizer plates.

ANCHOR RODS

L	A	B	C	D	T	Part No.	Equalizer Plate No.	Weight
84	1		2	5/8		GAR30	EYE	9
84	2			5/8	3/16	GAC303	EE	13
84	2			5/8	3/16	GAC305	EE	14
84	2	12	2-1/2	3/4	3/8	GAC3455TOP	EJ	25
120	2-1/2	12	3	1-1/4	1/2	GAC5655TOP	EJ	65
168	3	12	4	1-7/16	3/4	GAC5755TOP	EJ	125
192	4	12	6	1-1/4	1	GAC5855TOP	EJ	220
240	4	18	6	1-7/16	1	GAC5955TOP	EJ	310

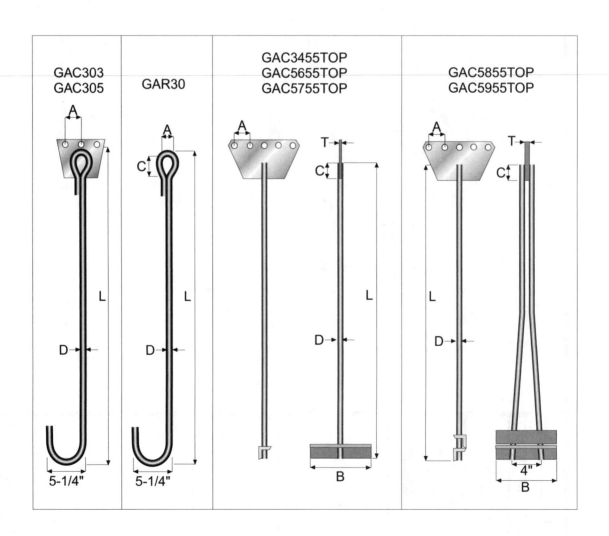

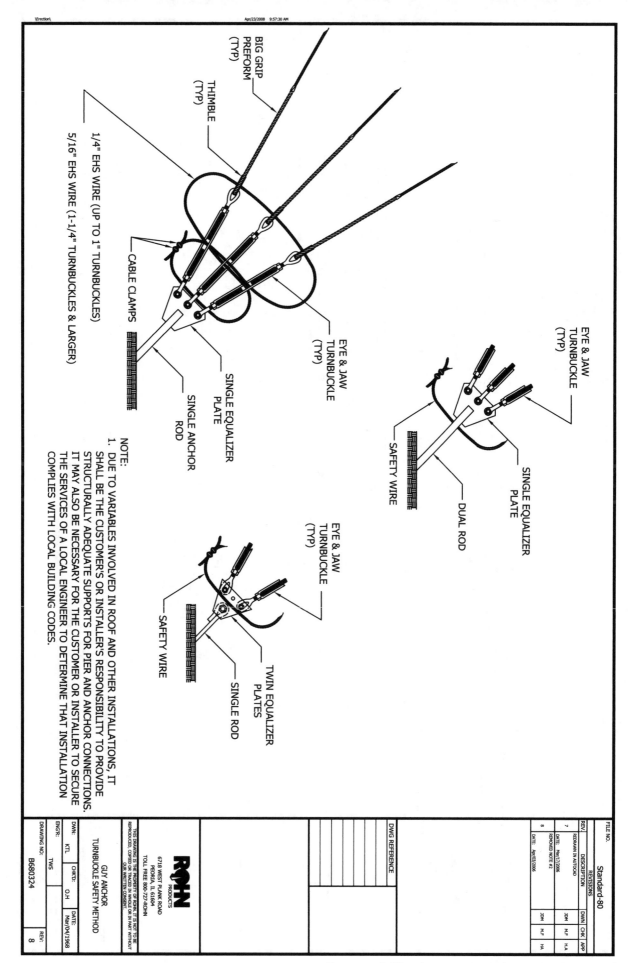

BIG GRIP
PREFORM
(TYP)

THIMBLE
(TYP)

1/4" EHS WIRE (UP TO 1" TURNBUCKLE)

5/16" EHS WIRE (1-1/4" TURNBUCKLES & LARGER)

CABLE CLAMPS

SINGLE EQUALIZER
PLATE

SINGLE ANCHOR
ROD

EYE & JAW
TURNBUCKLE
(TYP)

EYE & JAW
TURNBUCKLE
(TYP)

SAFETY WIRE

SINGLE EQUALIZER
PLATE

DUAL ROD

EYE & JAW
TURNBUCKLE
(TYP)

SAFETY WIRE

TWIN EQUALIZER
PLATES

SINGLE ROD

NOTE:
1. DUE TO VARIABLES INVOLVED IN ROOF AND OTHER INSTALLATIONS, IT
SHALL BE THE CUSTOMER'S OR INSTALLER'S RESPONSIBILITY TO PROVIDE
STRUCTURALLY ADEQUATE SUPPORTS FOR PIER AND ANCHOR CONNECTIONS.
IT MAY ALSO BE NECESSARY FOR THE CUSTOMER OR INSTALLER TO SECURE
THE SERVICES OF A LOCAL ENGINEER TO DETERMINE THAT INSTALLATION
COMPLIES WITH LOCAL BUILDING CODES.

ROHN
PRODUCTS

6718 WEST PLANK ROAD
PEORIA, IL 61604
TOLL FREE 800-727-ROHN

GUY ANCHOR
TURNBUCKLE SAFETY METHOD

DWG REFERENCE

FILE NO.

Standard-80

REV.	DESCRIPTION	DWN	CHK	APP
	REVISIONS			
7	REDRAWN IN AUTOCAD	JDM	H.F	H.A
	DATE: Mar/17/2006			
8	REMOVED NOTE #2	JDM	H.F	H.A
	DATE: Apr/03/2006			

DWN:	CHKD:	DATE:
KTL	O.H	Mar/04/1968

ENGR: TWS

DRAWING NO: B680324   REV: 8

Apr/23/2008   9:57:30 AM

\Erection\

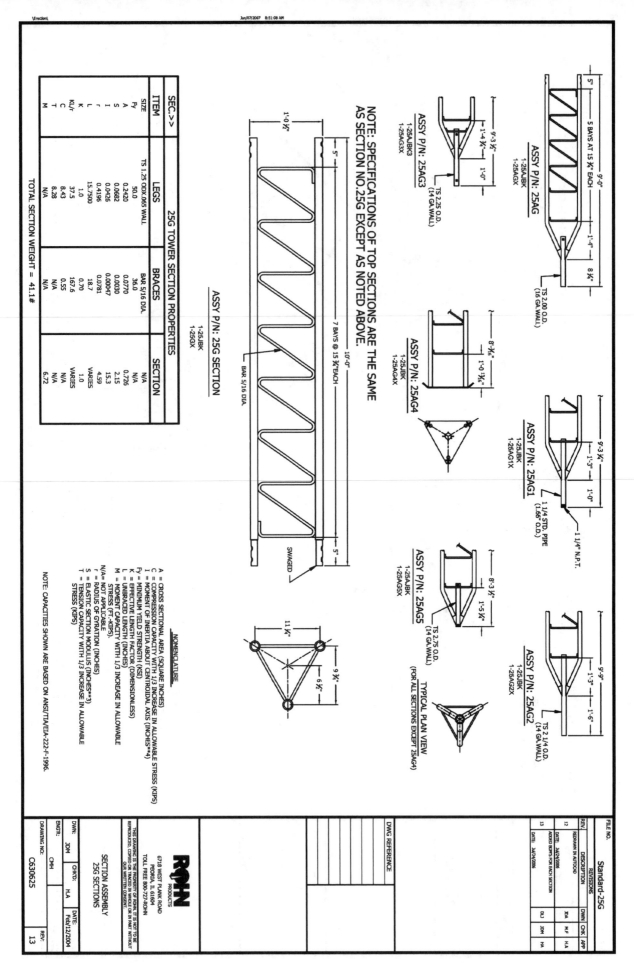

NOTE: SPECIFICATIONS OF TOP SECTIONS ARE THE SAME AS SECTION NO.25G EXCEPT AS NOTED ABOVE.

25G TOWER SECTION PROPERTIES

SEC.>>		25G
ITEM	LEGS	BRACES
		SECTION
SIZE	TS 1.25 ODX.065 WALL	BAR 5/16 DIA.
Fy	50.0	36.0
A	0.2420	0.0770
S	0.0682	0.0030
I	0.0426	0.00047
r	0.4196	0.0781
L	15.7500	VARIES
K	1.0	0.70
KL/r	37.5	167.6
C	8.43	0.55
T	8.28	N/A
M	N/A	N/A

	SECTION
	N/A
	0.726
	2.15
	15.3
	4.59
	18.7
	1.0
	VARIES
	N/A
	N/A
	6.72

TOTAL SECTION WEIGHT = 41.1#

ASSY P/N: 25G SECTION
1-25JBK
1-25GX

ASSY P/N: 25AG
1-25AJBK
1-25AGX

ASSY P/N: 25AG3
1-25AJBK3
1-25AG3X

ASSY P/N: 25AG4
1-25JBK
1-25AG4X

ASSY P/N: 25AG1
1-25JBK
1-25AG1X

ASSY P/N: 25AG5
1-25AJBK
1-25AG5X

ASSY P/N: 25AG2
1-25JBK
1-25AG2X

TYPICAL PLAN VIEW
(FOR ALL SECTIONS EXCEPT 25AG4)

NOMENCLATURE

A = CROSS SECTIONAL AREA (SQUARE INCHES)
C = COMPRESSION CAPACITY WITH 1/3 INCREASE IN ALLOWABLE STRESS (KIPS)
I = MOMENT OF INERTIA ABOUT CENTROIDAL AXIS (INCHES**4)
Fy = MINIMUM YIELD STRENGTH (KSI)
K = EFFECTIVE LENGTH FACTOR (DIMENSIONLESS)
L = UNBRACED LENGTH (INCHES)
M = MOMENT CAPACITY WITH 1/3 INCREASE IN ALLOWABLE STRESS (FT.-KIPS)
N/A= NOT APPLICABLE
r = RADIUS OF GYRATION (INCHES)
S = ELASTIC SECTION MODULUS (INCHES**3)
T = TENSION CAPACITY WITH 1/3 INCREASE IN ALLOWABLE STRESS (KIPS)

NOTE: CAPACITIES SHOWN ARE BASED ON ANSI/TIA/EIA-222-F-1996.

ROHN
PRODUCTS
6718 WEST PLANK ROAD
PEORIA, IL 61604
TOLL FREE 800-727-ROHN

THIS DRAWING IS THE PROPERTY OF ROHN. IT IS NOT TO BE REPRODUCED, COPIED OR TRACED IN WHOLE OR IN PART WITHOUT OUR WRITTEN CONSENT.

SECTION ASSEMBLY
25G SECTIONS

DWN: JDM   CHKD: H.A.   DATE: Feb/12/2004
ENGR:
DRAWN: CMH
DRAWING NO: C630625   REV: 13

FILE NO.	Standard-25G				
REV.	DESCRIPTION	DWN	CHK	APP	
	REVISIONS				
13	DATE: Jul/24/2006	DLJ	JDM	H.A	
12	ADDED BOLTS FOR EACH SECTION	JDA	M.F	H.A	
	DATE: Jul/24/2006				
REV.	REDRAWN IN AUTOCAD				

DWG REFERENCE

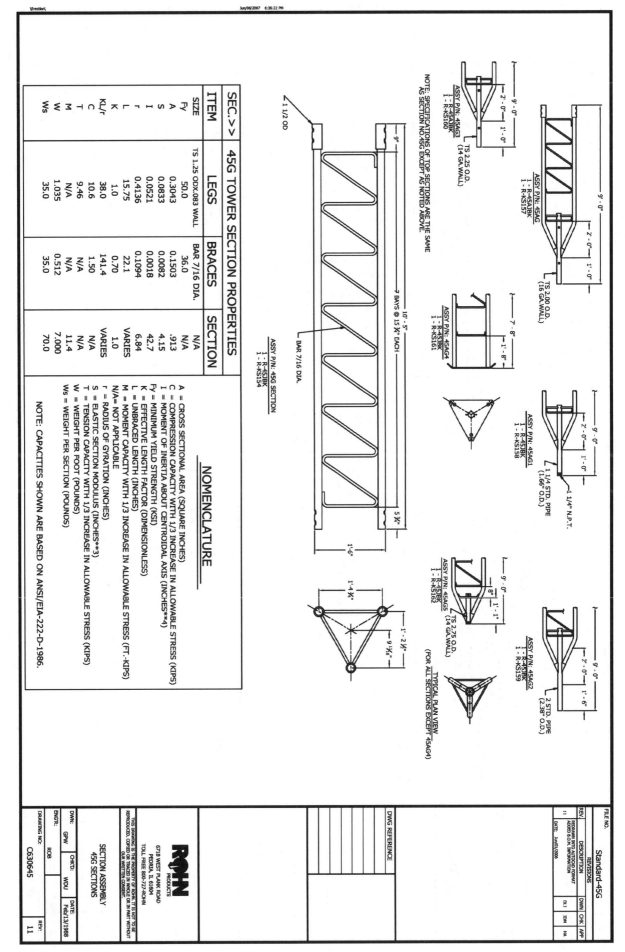

SEC.>> 45G TOWER SECTION PROPERTIES

ITEM		LEGS	BRACES	SECTION
SIZE		TS 1.25 ODX.083 WALL	BAR 7/16 DIA.	
Fy		50.0	36.0	N/A
A		0.3043	0.1503	N/A
S		0.0833	0.0082	.913
I		0.0521	0.0018	4.15
r		0.4136	0.1094	42.7
L		15.75	6.84	4.15
K		1.0	22.1	VARIES
KL/r		38.0	0.70	1.0
C		10.6	141.4	VARIES
T		9.46	1.50	N/A
M		N/A	N/A	N/A
W		1.035	0.512	11.4
Ws		35.0	35.0	0.512
				7.000
				70.0

NOTE: CAPACITIES SHOWN ARE BASED ON ANSI/EIA-222-D-1986.

NOMENCLATURE

A = CROSS SECTIONAL AREA (SQUARE INCHES)
C = COMPRESSION CAPACITY WITH 1/3 INCREASE IN ALLOWABLE STRESS (KIPS)
I = MOMENT OF INERTIA ABOUT CENTROIDAL AXIS (INCHES**4)
Fy = MINIMUM YIELD STRENGTH (KSI)
K = EFFECTIVE LENGTH FACTOR (DIMENSIONLESS)
L = UNBRACED LENGTH (INCHES)
M = MOMENT CAPACITY WITH 1/3 INCREASE IN ALLOWABLE STRESS (FT.-KIPS)
N/A = NOT APPLICABLE
r = RADIUS OF GYRATION (INCHES)
S = ELASTIC SECTION MODULUS (INCHES**3)
T = TENSION CAPACITY WITH 1/3 INCREASE IN ALLOWABLE STRESS (KIPS)
W = WEIGHT PER FOOT (POUNDS)
Ws = WEIGHT PER SECTION (POUNDS)

1 1/2 OD

BAR 7/16 DIA.

ASSY P/N: 45G SECTION
1 - R-45J3K
1 - R-KS154

7 BAYS @ 15¼" EACH

9"

10' - 5"

5¼"

1'-6"

NOTE: SPECIFICATIONS OF TOP SECTIONS ARE THE SAME
AS SECTION NO.45G EXCEPT AS NOTED ABOVE.

ASSY P/N: 45AG3
1 - R-45J3K
1 - R-KS160

TS 2.25 O.D.
(14 GA. WALL)

2' - 0"
9' - 0"
1' - 0"

ASSY P/N: 45AG
1 - R-45J3K
1 - R-KS157

TS 2.00 O.D.
(16 GA. WALL)

2' - 0"
9' - 0"
1' - 0"

ASSY P/N: 45AG4
1 - R-45J3K
1 - R-KS161

7' - 8"
1' - 8"

ASSY P/N: 45AG1
1 - R-45J3K
1 - R-KS158

1 1/4 STD. PIPE
(1.66" O.D.)
1 1/4" N.P.T.

2' - 0"
9' - 0"
1' - 0"

1'-4¾"
9 9/16"
1'- 2½"

TYPICAL PLAN VIEW
(FOR ALL SECTIONS EXCEPT 45AG4)

ASSY P/N: 45AG2
1 - R-45J3K
1 - R-KS162

8"
1'- 1"

ASSY P/N: 45AG5
1 - R-45J3K
1 - R-KS159

TS 2.75 O.D.
(14 GA. WALL)

2 STD. PIPE
(2.38" O.D.)

2' - 0"
9' - 0"
1' - 6"

ROHN PRODUCTS
6718 WEST PLANK ROAD
PEORIA, IL 61604
TOLL FREE 800-727-ROHN

THIS DRAWING IS THE PROPERTY OF ROHN. IT IS NOT TO BE
REPRODUCED, COPIED OR TRACED IN WHOLE OR IN PART WITHOUT
OUR WRITTEN CONSENT.

SECTION ASSEMBLY
45G SECTIONS

DWN: GPW | CHK'D: WDU | DATE: Feb/13/1988
ENGR: ROB
DRAWING NO: C630645 | REV: 11

FILE NO. Standard-45G

REVISIONS
REV	DESCRIPTION	DWN	CHK	APP
11	REDRAWN INTO AUTOCAD FORMAT, ADDED B.O.M. INFORMATION	DLJ	TDM	HA
	DATE: Jun/01/2006			

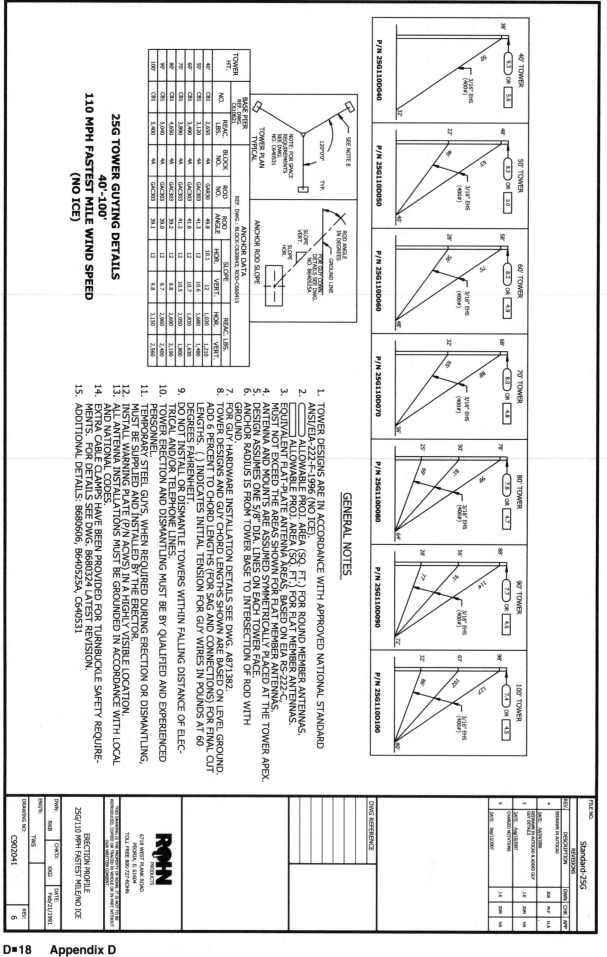

25G TOWER GUYING DETAILS
40'-100'
110 MPH FASTEST MILE WIND SPEED
(NO ICE)

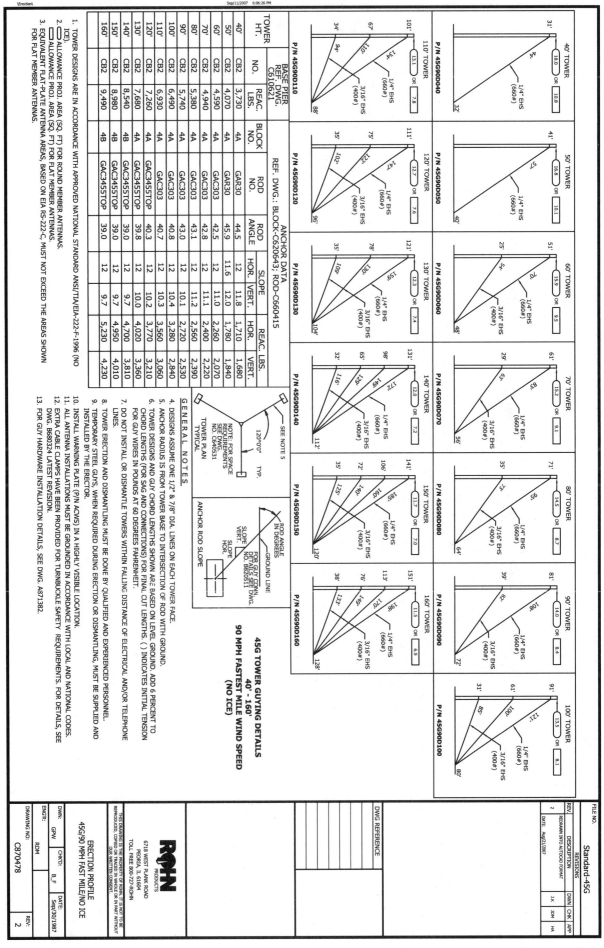

NOTES

NOTES

NOTES

NOTES

NOTES

NOTES

NOTES

NOTES

About the ARRL

The seed for Amateur Radio was planted in the 1890s, when Guglielmo Marconi began his experiments in wireless telegraphy. Soon he was joined by dozens, then hundreds, of others who were enthusiastic about sending and receiving messages through the air—some with a commercial interest, but others solely out of a love for this new communications medium. The United States government began licensing Amateur Radio operators in 1912.

By 1914, there were thousands of Amateur Radio operators—hams—in the United States. Hiram Percy Maxim, a leading Hartford, Connecticut inventor and industrialist, saw the need for an organization to band together this fledgling group of radio experimenters. In May 1914 he founded the American Radio Relay League (ARRL) to meet that need.

Today ARRL, with approximately 155,000 members, is the largest organization of radio amateurs in the United States. The ARRL is a not-for-profit organization that:

- promotes interest in Amateur Radio communications and experimentation
- represents US radio amateurs in legislative matters, and
- maintains fraternalism and a high standard of conduct among Amateur Radio operators.

At ARRL headquarters in the Hartford suburb of Newington, the staff helps serve the needs of members. ARRL is also International Secretariat for the International Amateur Radio Union, which is made up of similar societies in 150 countries around the world.

ARRL publishes the monthly journal *QST*, as well as newsletters and many publications covering all aspects of Amateur Radio. Its headquarters station, W1AW, transmits bulletins of interest to radio amateurs and Morse code practice sessions. The ARRL also coordinates an extensive field organization, which includes volunteers who provide technical information and other support services for radio amateurs as well as communications for public-service activities. In addition, ARRL represents US amateurs with the Federal Communications Commission and other government agencies in the US and abroad.

Membership in ARRL means much more than receiving *QST* each month. In addition to the services already described, ARRL offers membership services on a personal level, such as the Technical Information Service—where members can get answers by phone, email or the ARRL website, to all their technical and operating questions.

Full ARRL membership (available only to licensed radio amateurs) gives you a voice in how the affairs of the organization are governed. ARRL policy is set by a Board of Directors (one from each of 15 Divisions). Each year, one-third of the ARRL Board of Directors stands for election by the full members they represent. The day-to-day operation of ARRL HQ is managed by an Executive Vice President and his staff.

No matter what aspect of Amateur Radio attracts you, ARRL membership is relevant and important. There would be no Amateur Radio as we know it today were it not for the ARRL. We would be happy to welcome you as a member! (An Amateur Radio license is not required for Associate Membership.) For more information about ARRL and answers to any questions you may have about Amateur Radio, write or call:

ARRL—The national association for Amateur Radio
225 Main Street
Newington CT 06111-1494
Voice: 860-594-0200
Fax: 860-594-0259
E-mail: **hq@arrl.org**
Internet: **www.arrl.org/**

Prospective new amateurs call (toll-free):
800-32-NEW HAM (800-326-3942)
You can also contact us via e-mail at **newham@arrl.org**
or check out *ARRLWeb* at **www.arrl.org/**

FEEDBACK

Please use this form to give us your comments on this book and what you'd like to see in future editions, or e-mail us at **pubsfdbk@arrl.org** (publications feedback). If you use e-mail, please include your name, call, e-mail address and the book title, edition and printing in the body of your message. Also indicate whether or not you are an ARRL member.

Where did you purchase this book?   ☐ From ARRL directly   ☐ From an ARRL dealer

Is there a dealer who carries ARRL publications within:

☐ 5 miles   ☐ 15 miles   ☐ 30 miles   of your location?   ☐ Not sure.

License class:

☐ Novice   ☐ Technician   ☐ Technician with code   ☐ General   ☐ Advanced   ☐ Amateur Extra

Name_____   ARRL member? ☐ Yes ☐ No

_____   Call Sign _____

Address _____

City, State/Province, ZIP/Postal Code _____

Daytime Phone (   ) _____ Age _____

If licensed, how long? _____

Other hobbies _____   E-mail _____

_____

Occupation _____

For ARRL use only		ATRA
Edition	1 2 3 4 5 6 7 8 9 10 11 12	
Printing	1 2 3 4 5 6 7 8 9 10 11 12	

EDITOR, ANTENNA TOWERS FOR RADIO AMATEURS
ARRL—THE NATIONAL ASSOCIATION FOR AMATEUR RADIO
225 MAIN STREET
NEWINGTON CT 06111-1494

please fold and tape